Domestication of Plants
in the Old World

D1322190

Domestication of Plants in the Old World

The origin and spread of cultivated
plants in West Asia, Europe,
and the Nile Valley

SECOND EDITION

DANIEL ZOHARY

Professor of Genetics,
The Hebrew University of
Jerusalem, Israel

and

MARIA HOPF

Formerly Head of Botany Department,
Römisch–Germanisches Zentralmuseum,
Mainz, Germany

CLARENDON PRESS · OXFORD
1994

Oxford University Press, Walton Street, Oxford OX2 6DP
Oxford   New York
Athens   Auckland   Bangkok   Bombay
Calcutta   Cape Town   Dar es Salaam   Delhi
Florence   Hong Kong   Istanbul   Karachi
Kuala Lumpur   Madras   Madrid   Melbourne
Mexico City   Nairobi   Paris   Singapore
Taipei   Tokyo   Toronto
and associated companies in
Berlin Ibadan

Oxford is a trade mark of Oxford University Press

Published in the United States
by Oxford University Press Inc., New York

A catalogue record for this book is available from the British Library

Library of Congress Cataloging in Publication Data
Zohary, Daniel.
Domestication of plants in the Old World: the origin and spread
of cultivated plants in West Asia, Europe, and the Nile Valley/
Daniel Zohary and Maria Hopf. – 2nd ed.
Includes bibliographical references and index.
1. Agriculture, Prehistoric. 2. Plants, Cultivated–History.
I. Hopf, Maria. II. Title
GN799.A4Z64   1993   630'.93–dc20   93–10413

ISBN 0 19 854795 1 (Hbk)
ISBN 0 19 854896 6 (Pbk)

Printed in Great Britain by
Biddles Ltd, Guildford & King's Lynn

Preface to the second edition

Since the completion of the writing of the first edition in 1987, archaeo-botanical investigations and the study of the Old World's crops and their wild relatives continued, frequently at an accelerated pace. An impressive body of new evidence was added, both crop-wise and site-wise. Significantly, the new information does not contradict the main conclusions arrived at five or six years ago, but confirms them.

In this second edition, an attempt is made to integrate this new evidence. We have also filled a gap by adding a chapter on dye plants. The revision is most apparent in the vegetables, in the fruit trees, and in some of the minor grain crops. Only a few years ago, our knowledge of their origin and early history was embarrassingly fragmentary. At least for some of these crops, the evidence today permits a sounder synthesis.

Jerusalem   D.Z.
Mainz   M.H.
1993

Preface to the first edition

South-west Asia, Europe, and the Nile valley are unique today for the vast extent of archaeo-botanical exploration. In the last thirty years, hundreds of Mesolithic, Neolithic, and Bronze Age sites have been excavated in these territories. Plant remains in many sites have been expertly identified, culturally associated, and radiocarbon-dated, and the finds have given critical information on the plants that started agriculture in this part of the world.

Considerable progress has also been achieved in the field of the wild ancestry of Old World crops. The wild progenitors of most of these cultivated plants have now been satisfactorily identified, both by comparative morphology and by genetic analyses. The distribution and ecological ranges of the wild relatives have been established, and furthermore, comparisons between wild types and their cultivated counterparts have revealed the evolutionary changes which were brought about by domestication.

As a result of these achievements, south-west Asia, Europe, and Egypt emerge as the first major geographic area in the world in which the combined evidence from archaeology and the living plants permits a modern synthesis of crop plant evolution. The accumulated information provides reasonable answers to the following questions: (a) What were the first plants to be domesticated in the Old World? (b) Where can the earliest signs of their domestication be found? (c) What were the subsequent main developments in plant cultivation over these regions? (d) What crops have been introduced into this area from other parts of Asia and Africa? (e) When did all these events take place?

In the following chapters an attempt is made to answer these questions and provide a review of the origin and the spread of cultivated plants in south-west Asia, Europe, and Africa north of the Sahara, i.e. the classical 'Old World'. The aim was to trace plant domestication and crop plant evolution in this part of the globe from its early beginnings up to classical times. The treatment (Chapters 2–9) is crop by crop. Chapter 10 adds essential documentation on representative archaeological sites. The information given is based on work published up to 1985.

Jerusalem   D.Z.
Mainz   M.H.
1987

Contents

1
Sources of evidence for the origin and spread of cultivated plants

The study of the origin and spread of cultivated plants is an interdisciplinary venture based on evidence from numerous sources. Several disciplines, such as archaeology, botany, genetics, chemistry, anthropology, agronomy, and linguistics are involved (for review, see Harlan and de Wet 1973). Yet the different sources of evidence vary considerably in reliability and relative weight. The modern synthesis leans heavily on two principal sources:

(i) Information obtained through examination of plant remains retrieved from archaeological excavations.

(ii) Evidence gathered from living plants, particularly that on the wild ancestry of crops.

In both fields, the last thirty years have witnessed major discoveries which have radically changed our view on the origin of cultivated plants. They have transformed a realm of numerous speculations and few solid facts into a well-documented field.

In the light of this recent development the contributions of several classical tools had to be reconsidered critically. Some sources of evidence, such as linguistic comparisons, have retained their indicatory power; yet with the availability of direct evidence they carry much less weight. Other suppositions, such as Vavilov's (1949–50) equation of centres of diversity of cultivated plants with places of origin (for review, see Zohary 1970; Harlan 1971), proved to be incorrect and can no longer be used.

The following sections review the main sources of evidence on which the modern assessment of crop plant evolution is based. A list is given in Table 1.

Archaeological evidence

The primary contribution of archaeology to the understanding of crop plant evolution is by the recovery of plant remains in archaeological excavations, and by identifying what crop species they belong to. The accumulated evidence can contribute to answer the following questions:

(i) When and where do we find the earliest signs of cultivation of the examined crops?

(ii) How and when did the crops spread to attain their present distributions?

(iii) What were the early cultigens like? What were the main changes in the crops during domestication? Where and when did these changes take place?

Table 1. Sources of evidence on the origin and spread of cultivated plants

---

I. *Archaeological evidence*
  1. *Archaeoethnobotany*: Identification of plant remains retrieved from archaeological excavations in connection with cultural associations and ^{14}C-dating. Determination of the earliest signs of cultivation in these plants and their subsequent spread. Changes in crops in time and space. Crop assemblages in various cultures.
  2. *Additional evidence*
      (a) Artefacts: Evaluation of: (i) dated tools associated with cultivation, harvesting, and processing of crops; (ii) cultivation artefacts such as irrigation canals, terraces, lynchets, plough marks, and cultivation boundaries.
      (b) Art: Early drawings, paintings, and reliefs of cultivated plants.
      (c) Palynology: Appearance of pollen grains of crops and weeds in dated cores or site contexts.
      (d) Chemical analysis: Identification of crops by specific organic residues retained in ancient vessels, in charcoal, etc.

II. *Evidence from the living plants*
  1. *Search for the wild progenitors*: Identification of the nearest wild relatives of the cultivated crops by use of:
      (a) Comparative morphology and comparative anatomy (classical taxonomy).
      (b) Determination of genetic affinities by cytogenetic analysis and by protein and DNA resemblances.
  2. *Distribution and ecology of the wild progenitors*
      (a) Geographic distribution of the wild relatives (including weedy forms).
      (b) Characterization of the habitats and the main adaptations of the wild relatives.
  3. *Evolution under domestication*. Main trends of morphological, physiological and chemical changes. Patterns of variation under cultivation, particularly the distribution and adaptation of the principal cultivated races and their centres of diversity. Development of crop complexes (wild forms, weedy races, and cultigens). Methods of planting, maintenance, and usage.
  4. *Additional evidence*
      (a) Genetic systems: Characterization of the main systems operating under domestication. Especially: Reproductive systems (including vegetative propagation).
      (b) Genetic interconnections between cultivars and wild relatives.
      (c) Intentional and unconscious selections.

III. *Other pertinent sources*
  1. *Historical information*: Documentation in tablets, inscriptions, manuscripts, and books.
  2. *Linguistic comparisons*: Names of crops in various languages.
  3. *Circumstantial evidence*: Geological, hydrological, limnological, dendrochronological, anthropological, and zoological indications on the initiation and spread of agriculture.

Table 2. Preservation of plant remains in archaeological excavations

---

I. *Carbonized remains*:
  1. Charred during handling:
    (a) near a hearth/oven
    (b) in a drying kiln
    (c) in a storage pit/silo (when cleaned)
    (d) in pottery fired in a pottery kiln
  2. Charred by conflagration:
    (a) stored material
    (b) material embedded in daub, unfired bricks, and floors
    (c) thatching material
    (d) scattered or dumped material

II. *Plant impressions:*
  1. In pottery
  2. In bricks and daub

III. *Parched remains:*
  1. In arid regions:
    (a) in caves
    (b) in tombs and pyramids
    (c) in clay
  2. In temperate regions:
    (a) in sealed containers
    (b) in offerings embedded in walls

IV. *Waterlogged remains*:
  1. In lakes
  2. In bogs
  3. In wells
  4. In sites covered by rising seawater level

V. *Metal-oxide preservation:*
  1. Near silver
  2. Near copper or bronze
  3. Near iron

VI. *Petrified remains:*
  1. Siliceous mineralization
  2. Calcareous mineralization

Obviously the key to all answers is the availability of 'fossil evidence'; that is, sufficient amounts of dated, culturally defined, analysable plant remains. The following paragraphs survey the main conditions under which plant material survives in archaeological contexts. These are also listed in Table 2.

Carbonized remains

Charred remains are the commonest source of analysable plant material in archaeological excavations. Carbonization occurs upon exposure to high temperatures due to fires. Such 'baking' (under a limited supply of oxygen) converts the plant's organic compounds into charcoal. Since charcoal is not affected by bacteria, fungi, or other decomposing organisms, carbonized plant remains survive in most environments. This includes wet places where ordinary organic material decays rapidly. Carbonized plant remains in archaeological contexts are therefore not products of geological carbonization (true fossils). They represent only 'subfossil' elements baked by fire.

When charred slowly and mildly, wood, seed, nuts, and sometimes even fleshy fruits or whole ears of cereals, can still retain most of their morphological and anatomical features. The morphology and the microscopic anatomical structures are frequently preserved in astonishing clarity, which allows a reliable analysis of the plant remains.

At fairly high temperatures (between 200 and 400°C), carbonization causes characteristic deformations. In cereals, the most obvious changes are shrinkage in the length of the kernel together with a relative increase or 'puffing' in its circumference. Size reductions and specific patterns of swelling and/or cracking appear also after the charring of seed of flax, broad bean, pea, and several other grain crops. Moreover, some organs do not generally survive charring. Thus the seed coats in leguminous plants or the glumes and pales in cereals are only recovered on special occasions. Usually they disintegrate into powder. The intensity of the deformation depends, among other things, on the amount of water present in the seed (the drier the kernels, the less they are deformed), the spread of the heating, and the temperatures reached.

Substantial information on the effects of heating on the seed of various plants has been gained experimentally by simulation of charring in laboratory ovens. Grains of various cereals and seeds of several pulses and flax have been the main elements tested. A determination of the amount of shrinkage in the seed of various crops also provides a better idea of the actual life-size dimensions of charred seed discovered in excavations.

Carbonized plant material is recovered from digs either by direct collection or by separation techniques. There are lucky discoveries of hoards of burnt grains stored in containers or silos, which sometimes contain almost pure grains. In order to recover scattered remains embedded in site deposits, the excavator frequently resorts to separation by flotation. Water flotation is the simplest and cheapest technique, and usually separates the scattered charred remains present in the deposits effectively. In special cases flotation by heavier liquids (e.g. by carbon tetrachloride) is used.

Impressions on pottery, daub, and bricks

Imprints of grains and other plant parts on pottery are a means of documentation of crop plants in archaeological sites. Such imprints are found particularly on handmade vessels. Pottery is one of the main diagnostic objects in archaeology, and imprints on pottery therefore have an obvious advantage, since once detected, they can be culturally classified and dated. However, imprints are frequently pressed into gritty, rough pottery (the common type of ceramics in early periods). On such a background the print is rather blurred, and unequivocal interpretation of such findings is often difficult.

Another source of plant impressions is provided by daub and bricks. Straw, chaff, and similar dry plant material is often added to the wet clay to act as a tempering element. Plant parts can also become embedded in the clay by chance, and even if the organic matter does not survive well the impressions remain in the dried or fired clay. They can serve as negative moulds for casting and reproducing the former inclusions.

Parched plant remains

Preservation by desiccation, which blocks the processes of bacterial and fungal decomposition, occurs only under extreme dryness, so this source of evidence is confined almost entirely to very arid areas. Such parched remains can be of special importance because of their perfect preservation.

Outstandingly rich remains of dried plants have been discovered in Egypt. Here grains, fruits, vegetables, corms, and other parts of plants placed in pyramids and tombs give an excellent account of plant cultivation in the Nile valley during dynastic times. The finds include soft parts of vegetables, leaves, and flowers which hardly ever survive under other conditions. Several discoveries of parched material were also made in caves in the Dead Sea basin.

Waterlogged preservation

In Europe valuable information has been obtained by examining plant material sunk in peat bogs or buried in the mud at the bottom of lakes, seas, or wells. Anaerobic conditions in these environments (and the presence of humic acids in bogs) act as effective preservatives, and plant remains in such places frequently retain their most delicate features. Excellent examples of waterlogged preservation have been found at the Swiss lake-shore dwellings and in the stomach contents of human corpses retrieved from bogs in Denmark, Holland, and Germany.

Preservation by oxides of metals

Bronze, silver, and iron occasionally act as effective preservatives for plant material buried close to them. In humid situations they produce metal oxides which impregnate the plant remains. Because copper-, silver-, and iron-oxides are highly toxic to bacteria and fungi they block decomposition.

Mineralization

This type of preservation is brought about by filling of cell cavities by inorganic substances or by replacement of the content of cell walls by minerals. Most common is mineralization by calcium carbonate ($CaCO_3$) or by silica.

Seed coats and fruit shells of several plants undergo natural mineralization. Stones of hackberry (*Celtis*) contain, for example, large quantities of $CaCO_3$ and the nutlets of several Boraginaceae accumulate silica. They sometimes survive in archaeological deposits without further means of outside preservation. Also durable are silica structures (phytoliths) that characterize some types of grass cells.

Digested or partly digested remains

Preserved human faeces (coprolites) constitute an only partly exploited source of evidence. Because humans cannot digest cellulose, woody plant fragments and shelled seed frequently retain their features after passing through the alimentary tract. Therefore when faeces are charred, desiccated, or waterlogged, they often contain numerous identifiable plant fragments, which indicate the content of the human food in the tested culture. Coprolite examination has already contributed significantly to American environmental archaeology. In the Old World this source of evidence has not yet been exploited extensively, although some results are already available (Hillman 1986).

Chemical tests

Tar compounds present in charred plant remains and organic residues precipitated in ancient vessels can be identified by gas liquid chromatography, infra-red spectroscopy, and other tests used by organic chemists. Such detection is possible even when these substances survived in minute traces. Significantly, some of these chemical compounds are specific to a single crop species or a single plant product. They can be used as diagnostic traits for crop identification.

Evidence from the living plants

Several contributions to the understanding of the crop plant evolution are made by the study of the living plants.

A major contribution is the identification of the *wild progenitors* from which the various cultivated plants were derived. Once the wild ancestry has been determined the following examinations can be carried out:

(i) Comparison of the cultivated varieties ('cultivars') with their wild counterparts in order to determine the main morphological, physiological, chemical, and genetic changes that took place under domestication.

(ii) Assessment of changes in adaptation. Answers can be sought to the following questions: Which adaptations, that are vital under wild situations, have broken down under domestication? What are the new 'adaptive syndromes' that have evolved under domestication? Which selective forces are responsible for these changes? What genetic systems are involved?

(iii) Delimitation of the distribution areas of the wild progenitors. This often provides information on the place of origin of the crops.

A second major contribution comes from examination of the crops and the ways they are handled, particularly under traditional ('primitive') systems of agriculture. Such studies include:

(i) patterns of variation in each crop and their geographies;

(ii) methods of cultivation and use;

(iii) genetic systems operating in the various crops and breeding traditions used;

(iv) interconnections between the cultivated varieties and their wild forms, including the presence of weedy races.

The following paragraphs survey the main tools used for identification of wild progenitors. They also deal with some of the complications and problems involving the use of the wild relatives for elucidation of crop plant evolution.

Discovery of wild progenitors

As already mentioned, a principal goal in the study of the living plant is the identification of the wild progenitors of the crops, that is the wild stocks from which the cultivated plants evolved. Plant domestication is a relatively recent evolutionary event. Therefore, one can expect that most wild ancestors are still alive and include forms similar to those that existed in pre-agricultural times. Indeed, the wild progenitors of the majority of the world's main food plants have already been identified. Many of them became known only during the last twenty-five years.

Several complementary tests are available for the identification of the wild progenitors of crops. They all aim at determining which of the wild species, usually grouped together with the crop in the same biological genus, is closest to the cultivated plant.

(i) The *classical taxonomic approach* recognizes the wild progenitor by its close morphological resemblance to the crop. This is the oldest method. In some cases morphological comparison provides sound clues for the determination of ancestry. Yet many crops exhibit a bewildering morphological variation, very different from the patterns present in wild plants, and this can be rather confusing to relationship analysis. Critical evaluation in such cases necessitates genetic verification, which can be obtained through cytogenetic analysis and by comparative protein and DNA tests.

(ii) *Cytogenetic analysis* aims at elucidating the chromosomal affinities between the cultivated plant and the wild species. It also tests whether or not these wild taxa are separated from the crop (and isolated from one another) by hybrid sterility or other reproductive isolation barriers. Since domestication is a recent event, the crop and its wild progenitor should retain a considerable amount of homology in their chromosomes. In contrast, other species grouped in the genus were very probably formed long before the beginning of agriculture. They could have diverged considerably in their chromosomal constitution.

The principal tool of cytogenetics is a programme of crosses between species followed by examination of interspecific hybrids. Chromosome pairing in meiosis indicates the degree of chromosomal homology between the two parents. As a rule, the crop shows full homology and complete interfertility with only one of the wild species in the tested genus. Such a wild type is recognized as the ancestor of the crop. Together they comprise the 'primary' gene-pool of the crop. In contrast, other members of the genus are frequently chromosomally distinct and are separated from the crop by strong reproduction isolation barriers' such as cross-incompatibility, hybrid inviability, or hybrid sterility. Such species are often called 'alien species' and their chromosomes 'alien chromosomes'. They comprise the 'secondary' and 'tertiary' gene-pools of the crops (Harlan and De Wet 1971; Harlan 1975, p.111).

To summarize, fully fertile hybrids showing normal chromosome pairing in meiosis point to close genetic relationship between the tested parents and implicate the wild plant in the ancestry of the crop. Lack of chromosome homology and the presence of strong reproductive isolation barriers indicate long-established genetic divergence and rule out the tested wild plant from being a progenitor of the crop.

Chromosome analysis of cultivated plants has also to deal with complications due to *polyploidy*, i.e. the formation of new types (or even new species) by doubling of chromosome numbers. Evolution by polyploidy is common in the plant kingdom. Many wild plants (including progenitors of cultivated plants) are not standard diploids but polyploid entities. One class comprises auto-polyploids which increased their chromosome number from the standard of two dosages (diploid condition) to three sets (triploids), four sets (tetraploids), or even higher levels. Such increases are not uncommon among vegetatively propagated crops (corm and tuber plants, ornamentals and some fruit trees). A second class includes allo-polyploids, i.e. types formed by interspecific hybridization followed by chromosome doubling. This combines the genetic contents of two (or even more) donor species in a new hybrid species. Bread wheat is a product of such fusion under domestication (p.47). Cultivated tobacco and the New World cottons had a similar mode of origin. In such crops a special cytogenetic test known as 'genome analysis' helps to elucidate the polyploid origin and to identify the parental stocks which donated their chromosomes to the new polyploid entities.

(iii) Recent advances in *molecular biology* provide sensitive new tools for determination of genetic differences within and between species. These molecular techniques are increasingly used in evaluating genetic variation in crops and for solving problems in crop evolution.

Critical results have already been obtained from protein comparisons. Proteins are excellent genetic markers since they are the primary products of the genes. Differences in proteins therefore reflect differences in the hereditary material. Gel electrophoresis separation makes it possible today to discern variation and differences in numerous proteins. Indeed, in several crops examination of electrophoretically discernible seed storage proteins and allozyme systems have already added significantly to our understanding of genetic variation in crops and their wild relatives, and also to the elucidation of genetic distances between types and identification of progenitor stocks.

Developments in the use of restriction enzymes which cleave

DNA into identifiable fragments (RFLP or restriction fragment length polymorphism tests) and our growing ability to determine nucleotide sequences in chosen segments of DNA, have added powerful tools for estimating genetic distances between taxa. The impact of these new molecular techniques is starting to be felt in solving problems of the origin of cultivated plants (Gept 1993).

Distribution of the wild progenitors

The wild relatives can also frequently provide critical information about places of domestication. In many crops, the progenitors occupy limited geographic territories. Their distribution areas are much smaller than those of their cultivated derivatives. Because domestication is a recent development, it is safe to assume that the distributions of the wild forms (weeds excluded) have not undergone drastic changes since the beginning of cultivation. Delimitation of the wild relative's distribution thus marks the territory in which the crop could have been taken into cultivation. The narrower the distribution area, the more accurate the placement can be.

Fortunately the distribution of the wild progenitor of emmer wheat – a principal Old World 'founder crop' – is confined to the Near East 'arc' (Map 3, p.41). It is thus possible to plot the area where Neolithic agriculture could have started fairly accurately. The archaeological records have fully corroborated this supposition. The delimitation of the place of origin of the chickpea is even more precise. Its wild ancestor is endemic to the south-east part of Turkey (Map 9, p.104). Yet not all wild progenitors have such a limited distribution. Some (e.g. the wild relatives of the foxtail millet, oat, flax, and numerous fruit trees) are distributed over extensive territories. The use of their distributions for the determination of places of origin is much less accurate.

Weeds and domestication

Some crops seem to have entered cultivation not directly but by first evolving weed forms. The establishment of tilled fields gave an opportunity to numerous *unwanted* plants to invade the newly made habitats and to evolve as weeds. Weed evolution went hand in hand with crop cultivation and from the very start the control of these invaders seems to have been a major problem in agriculture. Yet noxious weeds are plants that have successfully adapted themselves to the ecology of the tilled ground. They are independent only because they retain their wild mode of seed dispersal, and germinate and develop in spite of the efforts of the cultivator to eradicate them. But if any of such weeds turns out to produce a valuable commodity it can eventually change its relationship with humans. The cultivator may follow the rule 'if you can't beat them, join them' and start to utilize the weed by intentionally planting its seed, harvesting its fruits, and selecting the better yielders. Several Old World crops are such 'secondary crops' i.e. plants that entered domestication through the back door of weed evolution (Vavilov 1949–50). They were added to the crop assemblage only after the firm establishment of the principal seed crops. Well-documented cases are those of the oat, *Avena sativa* (p.77), and of the false flax, *Camelina sativa* (p.132). Several other plants seem to have followed a similar evolution under domestication.

Classification and botanical names

Orientation in crop plant evolution is frequently complicated by inconsistencies in species delimitation and by proliferation of botanical names. As already noted, cultivated plants are, as a rule, very variable. Furthermore, evolution under domestication commonly involves drastic modifications to organs and traits that stay fairly uniform in wild plants. It is no wonder that traditional taxonomic treatments of crops suffered from over-splitting, since they were based almost entirely on morphological comparisons. Frequently, interfertile crop types were ranked as separate species and called by different botanical names because they looked so different. For example, classical cereal taxonomists recognized 12–15 species of cultivated wheats (see Table 3, p.24). Barley and common oat were each split into two or more species (Table 5, p.58). Similar splitting and species ranking characterized numerous other crops.

With the accumulation of cytogenetic information it has become increasingly clear that the traditional classification of many crops is inadequate and even misleading. Frequently two, three, or even half a dozen 'species' were found to be interfertile, chromosomally homologous, and genetically interconnected. Moreover, in many cases the conspicuous morphological distinctions turned out to be governed by single mutations. Ranking such types as independent species is unjustified. They represent only varieties within species and deserve only intraspecific ranking. In wheats, modern taxonomic revision has reduced the species number to four (Table 3). All cultivated barleys are grouped in a single species (Table 5), as are all common oats.

The discovery of the wild progenitors necessitated another nomenclature change. Because the wild plants and their cultivated derivatives are genetically interconnected, they too cannot be regarded as fully diverged species. According to internationally agreed taxonomic rules, once a wild ancestor is satisfactorily recognized, the crop and its wild relative cease to be treated as two separate species. Instead, they should be lumped in a single collective species, frequently also together with related weed types. In other words the wild and crop types are considered as subspecies or varieties of a single biological species and named accordingly.

Yet habits die hard. Old names and traditional classifications are still widely used by many workers. Wild progenitors, in particular, are commonly referred to as independent species. To avoid confusion, botanical orientation in crops should begin with the following questions:

(i) What are the main cultivated, weedy, and wild elements in the crop complex?

(ii) What botanical names are used by different workers for these intraspecific taxa of the crop complex?

(iii) What are the other fully divergent ('alien') species placed in the same genus?

Radiocarbon dating and dendrochronology

Radiocarbon (^{14}C) dating was developed by W.F. Libby at the University of Chicago soon after the Second World War and created a real breakthrough in archaeology (Gillespie 1986). Previously, one could date remains only by *relative* chronology based on stratigraphy and cultural associations. The introduction of ^{14}C tests brought about the *absolute* dating and made possible age comparisons between cultures in the various parts of the world.

Until several years ago, radiocarbon dating demanded large samples. Most tests were performed on charred wood or seed retrieved from marker beds. Since 1978, new methodologies have made ^{14}C dating possible using very small samples (e.g. a single charred grain), so that today one can *directly* date minute remains of crops. Such tests (radiocarbon accelerator mass spectrometry) are sometimes critical (Harris, 1986). They serve, among other things, to rule out errors of intrusion, i.e. the occasional translocation of plant remains from one bed to another as a result of boring by animals or some other interference.

Conventional radiocarbon dating is based on the assumption that the ratio between ^{12}C and ^{14}C in the atmosphere is even and remained constant throughout the ages. After incorporation into the organic material, the proportion of the radioactive isotope decreases due to its gradual disintegration. The accepted half-time of ^{14}C decay is 5568 years.

However, the assumption of atmospheric constancy of ^{14}C was found to be slightly incorrect. Some fluctuations seem to have occurred in the concentration of this isotope in the atmosphere in the past 12 000 years. This means that age estimates based on conventional ^{14}C time-scale are not fully calendar and are in need of some calibration. More precise dating was made possible by establishing the sequences of annual rings in wood remains of trees (oaks, bristlecone pine) for the last 9000 years and ^{14}C dating of the rings in these dendrological sequences. By such comparisons

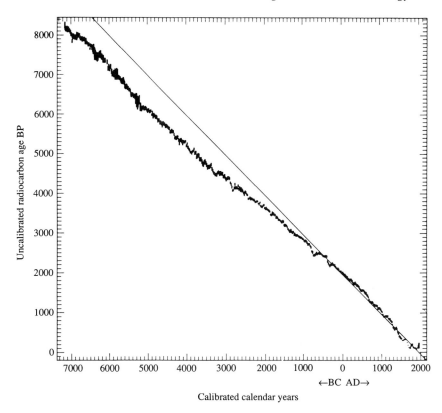

Fig. 1. The radiocarbon calibration curve for the past 9000 years based on Irish oak annual rings sequence (Pearson 1987). The straight line represents the ideal 1:1 correspondence between radiocarbon age and dendrological (calendar) age assuming constancy of ^{14}C concentration in the atmosphere. If conventional ^{14}C years were equivalent to calendar years, all of the data would fall on the diagonal line.

calibration curves have been constructed correlating radiocarbon dates with dendro- or calendar times. (Fig. 1). A continuous time-scale based on annual ring patterns and age sequences in wood remains is now available for the last 8000 years for bristlecone pine in California and for more than 9000 years for the oak in Ireland (Bowman 1990). Thus by means of dendrochronology, the radiocarbon time-scale can be calibrated against tree-ring chronology to calendar dates (Pearson 1987). The results indicate that for early archaeological contexts the ^{14}C time-scale represents somewhat reduced estimates. The corrected, calibrated ages of Neolithic samples are several hundred years older than those determined by the ^{14}C method. The relationship between the radiocarbon time-scale and the calendar dates is shown in Fig. 1. It shows the general slope of the curves back to 7000 BC. Note that at 5000 BC there is a divergence of about

800–900 years from the 1:1 relationship between uncalibrated radiocarbon age and calibrated calendar age.

As a rule, radiocarbon dates mentioned in this book represent *uncalibrated* dates.

The sign bc (in lower-case letters) used throughout this book, as well as BP (years before present, i.e. before 1950), denote *uncalibrated* ^{14}C dates. In contrast, historical dates and dendrologically *calibrated* radiocarbon dates (calendar years) are conventionally marked BC, AD, and calBP.

2
Cereals

Cereals (annual grasses cultivated for their grains) have been the principal crops of most civilizations. Their grains constitute the main source of calories for mankind. Before the discovery of America and the establishment of close contacts between Europe, the Far East, and Africa south of the Sahara, different parts of the world depended on cultivation of different staple cereals: Food production in the Mediterranean basin, Europe, the non-tropical parts of Asia, and (to some extent) the highlands of Ethiopia, was based primarily on wheat and barley. South and south-east Asia had rice, America maize, and Africa south of the Sahara sorghum, pearl millet, and several other endemic grasses.

Cereals thrive in open ground (the cultivated field) and complete their life cycle in less than a year. The nutritive value of their grains is generally high, and the seed can be stored for long periods. In most cereals the kernels are packed with starch, and in some, such as in wheats or oats, they also contain an appreciable amount of protein. Compared with grain crops belonging to other plant families (e.g. legumes), yields in cereals are relatively high. But they depend heavily on soil fertility and the availability of fixed nitrogen.

Wheat and barley have been the traditional staples of Europe and west Asia, and they are also the principal 'founder crops' that started food production in this part of the world. The first definite signs of wheat and barley cultivation appeared in the Near East 'arc' in the second half of the 8th millennium bc. Later, grains of these cereals constituted the bulk of plant remains retrieved from Near East Neolithic and Bronze Age contexts. Wheat and barley were also the main domesticates that made possible the explosive expansion of the Neolithic agriculture from its 'core area' to the vast territories of west Asia, Europe, and North Africa – from the Atlantic coast to the Indian subcontinent and from Scandinavia to the Nile Valley.

Several other grasses entered agriculture in west Asia and Europe, but apparently only after the firm establishment of the 'first wave' in Near-Eastern cereals. Two early domesticates were rye and common millet. They were followed, much later, by oats. Sorghum and rice were introduced to the Near East and the Mediterranean basin only in classical times, after their domestication in Africa (sorghum) and south Asia (rice).

Wheat, barley, and in fact many other prominent cereals differ from the majority of flowering plants in their pollination behaviour. While the great majority of the world's plant species are allogamous or cross-pollinated, most cultivated seed crops are self-pollinated. Significantly, selfing is also a characteristic feature of their wild progenitors. In wheat, barley, and most other grasses it is brought about by precocious shedding of the pollen inside the florets; that is, before their opening.

Was it a mere chance that the first plants that were successfully domesticated in the Old World were selfers? Several facts suggest that self-pollinated plants were better suited to domestication than cross-pollinated candidates. One major advantage of self- over cross-pollination in incipient domesticates is the fact that selfing isolates the crop reproductively from its wild progenitor. It enables the farmer to grow a desirable cultivar in the same area in which its wild relatives abound, without endangering the identity of the cultivar by genetic swamping. If indeed domestication of the Near East cereals took place in areas where the wild progenitors were common, genetic separation between tame and wild was a major asset. Under cross-pollination the initial small patches of cultivation would have been exposed to large quantities of pollen from wild-growing relatives. Safeguarding the identity of the cultivated varieties would have become difficult or even impossible.

A second advantage of self-pollination lies in the genetic structure maintained within the crop. Selfing results in splitting the crop's gene-pool into independent homozygous lines. Variation is thus structured in the form of numerous true breeding cultivars. Since they are automatically 'fixed' by the pollination system, they can easily be maintained by the farmer, even when planted together. In contrast, the preservation of varietal identity in cross-pollinated plants is much more problematic. It requires repeated selection towards the desired norms, constant care to avoid the mixing of types and the prevention of contamination from undesirable plants. It is therefore not surprising that early successes in plant domestication involve selfers. In fact, the 'first wave' domesticates in the Near East, namely emmer wheat, einkorn wheat, barley, pea, lentil, chickpea, bitter vetch, and flax, are all self-pollinators. Cross-pollinated crops entered agriculture later and comprise only a small minority among the traditional grain crops. Rye and broad bean are the only important cross-pollinated grain crops of the Old World.

It is noteworthy that wheats, barley, and all other 'first wave' Near-Eastern seed crops, are not obligatory selfers but predominantly self-pollinated plants, in which also rare events of cross-pollination occur. Such a pollination system is admirably suited for rapid crop evolution because it produces new types recurrently. Occasional cross-pollination provides the crop with genetic flexibility and serves to combine and reshuffle genes from different sources. The numerous cycles of inbreeding that follow the rare events of crossing lead to the fixation of many new recombinant lines. The more attractive ones among them can easily be picked up by the grower.

Cereal cultivation is based on: sowing the seed in the tilled field; reaping the mature spikes or panicles; threshing of the grains. The introduction of these practices automatically initiated numerous changes in grasses taken into cultivation, setting them apart from their wild progenitors (Harlan *et al.* 1973; De Wet 1975). Most conspicuous is the unconscious selection for types in which the mature seed is retained on the mother plant, and in which the wild-type adaptation for seed dissemination has broken down (Zohary 1969). In cereals this implies a shift from shattering spikes or panicles in wild forms to non-shattering types in plants under domestication. In the wild, survival depends on seed dispersal and the various wild cereals have evolved elaborate dissemination devices. In wild wheats (pp.32, 40) and wild barley (p.60) the seed-dispersal unit (diaspore) comprises a single internode of the ear. Disarticulation of the spike's rachis at each segment is thus an essential element of the wild-type dispersal. Plants in wild populations are constantly selected for quick shattering of their mature ears. In contrast, the introduction of planting and harvesting brings about unconscious, automatic selection in exactly the opposite direction. Under the new system, a sizable proportion of the seed produced by brittle plants will shatter and would not be included in the harvest, while all grains produced by non-shattering mutants 'wait on the stalks' to be reaped by the grower. Under cultivation, non-shattering individuals have therefore a much better chance to contribute their seed to the subsequent sowing. All in all, non-brittle (or less brittle) mutants which were totally disadvantaged under wild conditions become highly successful under the new system. Thus, when wild cereals are brought into cultivation, one should expect selection for non-brittle forms whether or not the cultivator is aware of this trait. Furthermore, the incorporation of non-brittle mutants makes the crop fully dependent on humans, as non-shattering plants lose their seed dispersal ability, and can no longer survive under wild conditions.

The establishment of non-shattering mutants in cereal crops was very probably a fast process. Genetic considerations indicate that such a shift could have been accomplished in the course of a few dozen generations (Hillman and Davies 1990).

A second major outcome of introducing wild cereals into the regime of cultivation is the breakdown of the wild mode of seed germination. Most wild grasses, especially annuals growing in Mediterranean-type or desert climates, depend for their survival on regulation of germination. A common adaptation is a delay of germination from the time of seed maturation until the onset of favourable conditions in the next growing season. In the Mediterranean basin this means germination inhibition – even if the seed is wetted – from the time of seed maturation in early summer to the start of the rainy season in the following autumn. Another adaptation is the prolongation of germination over two or more years. This protects the population from the crippling effects of dry or otherwise bad years. In wild emmer wheat the dispersal unit or diaspore contains two kernels. One germinates in the ensuing autumn; while in the second kernel germination is frequently delayed until the following year. Wild barley diaspores each contain only a single seed, but not all dispersal units germinate in the first winter. Under cultivation, this wild-type regulation of germination is no longer advantageous, and mutants in which germination inhibition has broken down are selected automatically. This type of selection has evidently been operative in the Old World cultivated cereals. Most cultivars have lost the germination inhibition patterns which characterized their wild progenitors, so that all their seeds germinate at the same time whenever planted by the farmer.

Several other traits which characterize cultivated cereals (the 'domestication syndrome' traits) seem to be the outcome of unconscious selection under domestication (for review see Harlan *et al.* 1973). They include:

(i) selection towards erect types, synchronous tillering and uniform ripening;

(ii) increase of seed production by addition of fertile florets and increase in the size of the inflorescence or the number of ears or panicles produced per plant;

(iii) reduction of awns, glume thickness, and grain investment.

Wheats: *Triticum*

Wheats are the universal cereals of Old World agriculture. Together with barley, they constituted the principal grain stock that founded Neolithic agriculture and the main element responsible for its successful spread. Since this early start, wheats have retained their crucial role in Old World food production and have given rise to numerous advanced types. Today, wheats rank first in the world's grain production and account for more than 20 per cent of the total food calories consumed by humans. They are now extensively grown throughout the temperate, Mediterranean-type, and subtropical parts of both hemispheres of the world.

Wheats are superior to most other cereals (e.g. maize, rice, or barley) in their nutritive value. Their grains contain not only starch (carbohydrate content of wheat grains is 60–80 per cent), but also significant amounts of protein (8–14 per cent). The gluten proteins present in the seed endosperm give wheat dough its stickiness, and its ability to rise when leavened, in other words, unique baking qualities. Wheats were, and still are, the preferred staple food of traditional farming communities throughout the Old World from the Atlantic coast of Europe to the northern parts of the Indian subcontinent, and from Scandinavia and Russia to Egypt. Thus it is not surprising that in numerous cultures food has been equated with bread.

Several distinct species of the genus *Triticum* L. were brought into cultivation. Practically all modern wheat cultivars belong to two species: (i) bread wheat, *Triticum aestivum*, with $2n = 42$ chromosomes, valued for the baking of high rising bread, and (ii) hard or durum-type wheat, *T. turgidum*, with $2n = 28$ chromosomes, used for the preparation of macaroni and low rising bread. Other wheat species, as well as more primitive forms of the above-mentioned two species, were important in the past and survive today only as relics. All wheats are almost fully self-pollinated. Because of this fact, variation in these cereals is moulded in the form of numerous inbred lines.

Chromosomally wheats comprise a polyploid series (for review of wheat cytogenetics, see Riley 1965; Sears 1969; Feldman 1976). Some cultivated forms have a diploid chromosome number ($2n = 14$) and contain two sets of a single genome (designated AA). Other wheats are tetraploid ($2n = 28$ chromosomes) and combine two distinct genomes (either AABB or AAGG). Still others are hexaploid ($2n = 42$ chromosomes) and contain three different genomes (AABBDD). Thus wheats fall into four cytogenetic groups: one diploid, two tetraploid, and one hexaploid. Forms within each group are interfertile and share the same chromosome constitution. In contrast, hybrids between groups are highly sterile. The modern classification of cultivated wheats and their closely related wild types (MacKey 1966) is based also on these cytogenetic criteria. The following four principal species are recognized today in the genus *Triticum*.* Three species, the diploid and the two tetraploids, contain both cultivated varieties and wild counterparts. The fourth species is a hexaploid and contains only cultivated forms.

1. Diploid *T. monococcum* L. or einkorn wheat (genomic designation AA) comprises both wild and cultivated forms (Fig. 2). Cultivated einkorn with its characteristic hulled grains was an important grain crop in the past. It survives today only as a relic.

2. Tetraploid *T. turgidum* L. (genomic designation AABB) comprises wild emmer wheat, cultivated emmer wheat, durum wheat, and several other cultivated tetraploid forms (Figs 3 and 4). From the start of agriculture emmer emerges as the principal stock of wheat. It gave rise

---

* Excluding the closely related *Aegilops* L., which comprises some 20 species. Some *Aegilops* species show close cytogenetic links with the wheats. Indeed diploid *Aegilops* species participated in the formation of the polyploid *Triticum* species. For these reasons several workers (see, for example, Morris and Sears 1967) lump both genera and include the *Aegilops* species within *Triticum*. Others (e.g. Zohary 1965) keep the traditional classification but consider *Triticum* and *Aegilops* as a single natural group (the 'wheat group').

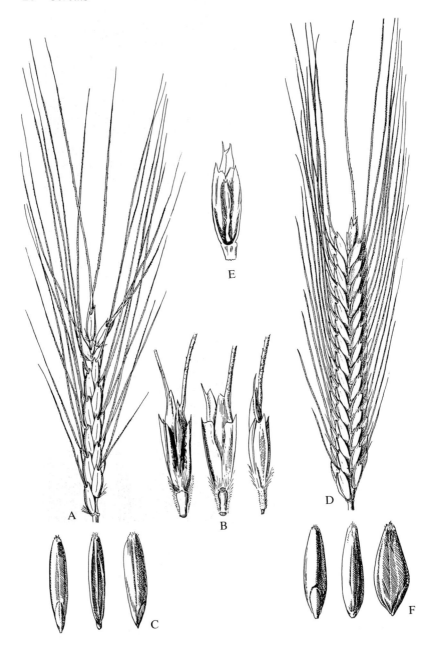

Fig. 2. Diploid einkorn wheats, *Triticum monococcum*. Left: A – ear (1:1), B – spikelet (2:1), and C – grain (3:1) of wild einkorn, *T. monococcum* subsp. *boeoticum*. Right: D – ear (1:1), E – spikelet (2:1), and F – grain (3:1) of cultivated einkorn, *T. monococcum* subsp. *monococcum*. (Schiemann 1948.)

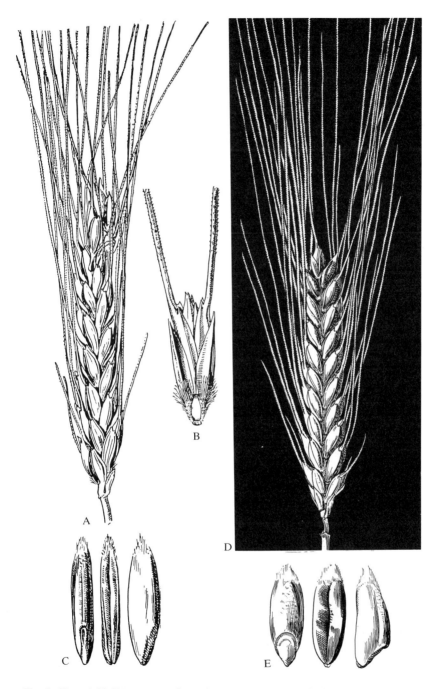

Fig. 3. Tetraploid *Triticum turgidum* wheats. Left: A – ear (1:1), B – spikelet (2:1), and C – grain (3:1) of wild emmer wheat, *T. turgidum* subsp. *dicoccoides*. Right: D – ear (1:1), and E – grain (3:1) of cultivated emmer wheat, *T. turgidum* subsp. *dicoccum*. (Schiemann 1948.)

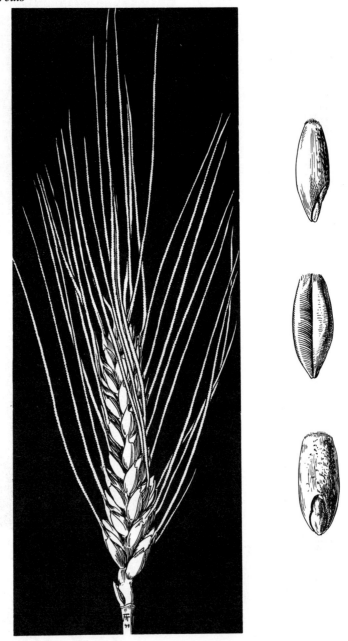

Fig. 4. Tetraploid *Triticum turgidum* wheats. Left: ear (1:1) and grain (3:1) of free-threshing durum wheat, *T. turgidum* subsp. *durum*. Right: ear (1:1) and grain (3:1) of free-threshing rivet wheat, *T. turgidum* subsp. *turgidum*. (Schiemann 1948.)

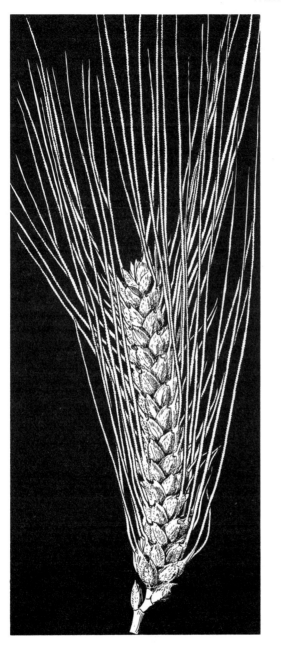

to the wide range of present-day, free-threshing, tetraploid durum-type wheats. By hybridization it formed an additional species: the hexaploid bread wheat (see below).

3. Tetraploid *T. timopheevi* Zhuk. (genomic designation AAGG) includes both wild and cultivated hulled forms. Cultivated Timopheev's wheat is endemic to a small area in Georgia, and seems to represent only a local episode in wheat-crop evolution.

4. Hexaploid *T. aestivum* L. or bread wheat (genomic designation

Table 3. Classification of wheats, *Triticum* L.: Main morphological types or species according to traditional classification and their modern grouping on the basis of cytogenetic affinities

Traditional classification	Modern grouping
1. Wild einkorn *T. boeoticum* Boiss. emend. Schiem. (brittle, hulled) including single grain forms: *T. aegilopoides* Link Bal.; and two grain forms: *T. thaoudar* Reuter and *T. urartu* Tuman. 2. Cultivated einkorn *T. monococcum* L. (non-brittle, hulled)	(i) Diploid ($2n = 14$) einkorn wheat Genomic constitution: AA Both wild and cultivated forms Collective name: *T. monococcum* L.
1. Wild emmer *T. dicoccoides* (Körn.) Aarons. (brittle, hulled) 2. Cultivated emmer *T. Dicoccum* Schübl. (non-brittle, hulled) 3. Macaroni or hard wheat *T. durum* Desf. (cultivated, free-threshing) 4. Rivet wheat *T. turgidum* L. (cultivated, free-threshing) 5. Polish wheat *T. polonicum* L. (cultivated, free-threshing) 6. *T. carthlicum* Nevski (= *T. persicum* Vav.) (cultivated, free-threshing) 7. *T. parvicoccum* Kislev (cultivated, free-threshing) small grained archaeobotanical forms	(ii) Tetraploid ($2n = 28$) emmer, durum, etc. Genomic constitution: AABB Both wild and cultivated forms Collective name: *T. turgidum* L.
1. Wild Timopheev's wheat *T. araraticum* Jakubz. (brittle, hulled) 2. Cultivated Timopheev's wheat *T. timopheevi* Zhuk. (non-brittle, hulled)	(iii) Tetraploid ($2n = 28$) Timopheev's wheat Genomic constitution: AAGG Both wild and cultivated forms Collective name: *T. timopheevi* Zhuk.
1. Spelta *T. spelta* L. (hulled) 2. *T. macha* Dekr. & Men. (hulled) 3. *T. vavilovii* Tuman. (hulled) 4. Bread wheat *T. aestivum* L. (= *T. Vulgare* Host; *T. sativum* Lam.) (free-threshing) 5. Club wheat *T. compactum* Host. (= *T. aestivo-compactum* Schiem.) (free-threshing) 6. Indian dwarf wheat *T. sphaerococcum* Perc. (free-threshing)	(iv) Hexaploid ($2n = 42$) bread wheat Genomic constitution: AAGG Only cultivated forms Collective name: *T. aestivum* L.

AABBDD) originated under cultivation by the addition of the DD chromosome complement of the wild grass *Aegilops squarrosa* L. to the tetraploid AABB *turgidum* wheats. The extraordinarily variable *T. aestivum* group constitutes the most important wheat crop of today (Fig. 5). It comprises several primitive hulled types (spelta wheat) and numerous free-threshing forms (including modern bread wheat).

Traditional wheat classification has suffered from excessive splitting. Until recently, wheat students regarded every main morphological type in the cultivated wheats as an independent species and recognized more than a dozen distinct species in the genus *Triticum* (for details see the treatments of Percival 1921; Schiemann 1948; Zhukovsky 1964). But in view of the available information on genetic affinities between types, such a species delimitation is no longer justified. Most wheat investigators today group all wheats in four biological species. Table 3 lists the various traditional species names in *Triticum* together with the modern grouping of the wheats. Table 4 presents the wild stocks from which the cultivated cereals were derived and summarizes the main evolutionary events that led to the formation of the crops.

Cultivated wheats fall into two distinct classes according to their response to threshing. The more primitive forms, i.e. diploid einkorn, tetra-ploid emmer, and hexaploid spelta have *hulled* grains. Their kernels are invested by tough pales and spikelet glumes. Consequently, the products of threshing are spikelets, not grains (Fig. 6). More advanced cultivated wheats, i.e. tetraploid durum-type and hexaploid bread wheats, are *free-threshing*. Their glumes and pales are thinner and do not invest the grains tightly. Threshing releases the naked kernels. The handling of hulled wheats is therefore different from that of the free-threshing ones. In the first, the spikelets are stored and marketed. Before their use the grains of hulled wheats have to be freed – usually by pounding. The utilization of naked wheats is simpler. After threshing, the free grains are winnowed and stored ready for milling. Because of the different appearance of the marketed products, hulled and free-threshing wheats were often regarded in antiquity as different cereals and called by different names. Yet one has to bear in mind that hulled and naked wheat forms can be genetically very close to one another so that they belong to the same species (Table 3). In present-day *T. aestivum*, the difference between hulled and free-threshing varieties is governed mainly by a single mutation (the q gene). In contrast, in most free-threshing *T. turgidum* wheats, the shift from 'hulledness' to 'nakedness' was brought about by a polygenic system.

Another important morphological trait in wheats is the manner in which the ear shatters in wild forms, or fractures in the cultivated forms. Wild wheats are adapted to disseminate their seed by having brittle ears that disarticulate at maturity into individual spikelets. In wild

Fig. 5. Hexaploid *Triticum aestivum* wheats. A – Ear and grain of spelta wheat. *T. aestivum* subsp. *spelta*. B – Ear and grain of club wheat. *T. aestivum* subsp. *compactum*. C + D – Ears and grains of awned and awnless varieties of bread wheat, *T. aestivum* subsp. *aestivum*. Ears 1:1; grains 3:1. (Schiemann 1948.)

einkorn (genome AA; Fig. 2), wild emmer (AABB; Fig. 3), and wild Timopheev's wheat (AAGG), the point of disarticulation is *below* each spikelet. Each spikelet with the wedge-shaped rachis internode at its base constitutes an arrow-like device that inserts the seed into the ground (Zohary 1969). In *Aegilops squarrosa* (genome DD), the disarticulation point is *above* the spikelet. The dispersal unit is a 'barrel-shaped' spikelet with a characteristic rachis internode joined to its inner side (Fig. 13). In contrast, all cultivated wheats have non-brittle ears that stay intact after maturation. They thus depend on us for their reaping, threshing and sowing. Yet threshing fractures the ears of non-brittle wheats in different ways, and significantly, these differences are also discernible in archaeological remains.

(i)   The ears of hulled diploid and tetraploid cultivated wheats (einkorn, emmer, and Timopheev's wheat) break on threshing at the same point at which their wild counterparts disarticulate spontaneously. In other words, threshing in these still primitive wheats 'mimics' the shattering pattern of their wild progenitors. The individual spikelet with the internode at its base is the product of threshing (Fig. 6, A and B).

(ii)   The rachis of hulled hexaploid spelta-type wheats (AABBDD) has two weak points at each spikelet – one inherited from the AABB progenitor and the second endowed by the DD contributor. Four different combinations of fracturing are possible. The diagnostic fracturing arrangement is that in which the rachis internode remains attached to the inner side of the spikelet (Fig. 6C). In archaeological digs, spikelets which follow the disarticulation pattern of *Ae.*

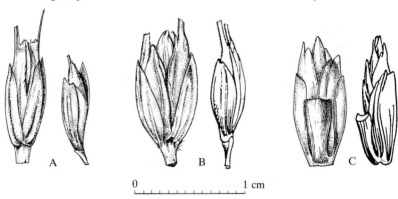

0 _____ 1 cm

Fig. 6. The threshing products of the three main types of cultivated hulled wheats: A – Einkorn, *Triticum monococcum* subsp. *monococcum*. B – Emmer, *T. turgidum* subsp. *dicoccum*. C – Spelta. *T. aestivum* subsp. *spelta*. (Modern material.)

Table 4. The wild progenitors of the main cultivated wheats and their principal derivatives under domestication

Wild einkorn wheat *T. monococcum* subsp. *boeoticum*. Diploid (AA). Brittle ears, hulled grain.

→ domestication →

Cultivated einkorn wheat *T. monococcum* subsp. *monococcum*. Diploid (AA). Non-brittle ears, hulled. Widely cultivated in the past; relic today.

Wild emmer wheat *T. turgidum* subsp. *dicoccoides*. Tetraploid (AABB). Brittle ears, hulled grain.

→ domestication →

Cultivated *T. turgidum*. Tetraploid (AABB). Non-brittle ears. First: hulled emmer wheat *T. turgidum* subsp. *dicoccum* (widely distributed in the past; relic today). Directly derived from it: free-threshing forms, mostly *durum* wheat (common in Mediterranean-type climates)

Wild *Aegilops squarrosa*. Diploid (DD). Never domesticated; but contributed its genome to hexaploid wheats.

addition of the D genome through hybridization

Cultivated *T. aestivum*. Hexaploid (AABBDD). Non-brittle. First: hulled *spelta*-type forms (relic today). Directly derived from them: free-threshing *bread wheats* (most common and most variable wheats today).

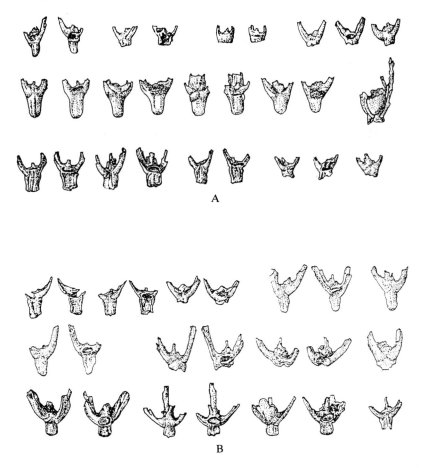

Fig. 7. Glume forklets, the diagnostic elements for the recognition of hulled wheats in archaeological remains. A – Einkorn. B – Emmer. Late Neolithic Goljamo Delcevo, Bulgaria. (Hopf 1975.)

squarrosa constitute a reliable proof for the presence of hexaploid spelta-type wheats.

(iii) In free-threshing wheats (tetraploid AABB durum-type and hexaploid AABBDD bread wheats) the rachis of the ear is thickened throughout its length. Threshing breaks the thinner glumes and pales at or near their base, and releases the naked grains. The thickened rachis breaks irregularly, frequently into segments of two to five internodes (Fig. 8).

Charred kernels and/or spikelets of hulled wheats and carbonized grains of free-threshing wheats constitute the bulk of wheat remains in archaeological

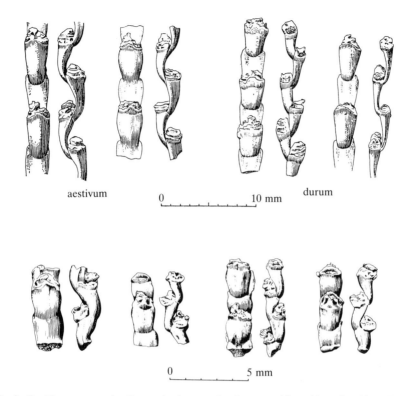

aestivum   0 ⌞————⌟ 10 mm   durum

0 ⌞————⌟ 5 mm

Fig. 8. Rachis segments, the diagnostic elements for the recognition of free-threshing wheats in archaeological remains. Upper row: Rachis fragments separated from among threshed grains of modern bread wheat (left) and durum wheat (right). Lower row: Carbonized rachis fragments of free-threshing wheats. Neolithic Tell Ramad, Syria. (van Zeist 1976.)

digs. In some sites, prints on clay (pottery, bricks, daub, floors) were detected. There are also few finds of whole ears and/or spikelets of dried 'mummified' material. The latter come from Egypt and other extremely dry locations.

In the hulled wheats, burning frequently pulverizes and destroys the upper parts of the investing glumes and pales, exposing and freeing the grains. But the bases of the spikelets often survive, and accompany the grains as discrete units. These 'glume forklets' (Fig. 7) are tell-tale indicators for the identification of hulled wheats in archaeological contexts.

In free-threshing wheats, the spikelet elements are winnowed out. They do not appear in the grain finds. But some of the fragments of the tough rachis (particularly the more basal parts) are similar to the kernels in their size and weight; and upon winnowing they occasionally stay with the grains. These fragments usually char well and become critical markers (Fig. 8) for the presence of free-threshing wheats in archaeological digs (van Zeist 1976; Hillman 1984).

Einkorn wheat: *Triticum monococcum*

Einkorn is a relatively uniform, diploid ($2n = 14$ chromosomes) wheat with characteristic hulled grains and delicate ears and spikelets (Fig. 2). Most cultivated einkorn varieties produce one grain per spikelet, hence its name, but varieties with two grains exist as well. Today einkorn is a relic crop, although it is still sporadically grown in western Turkey, the Balkan countries, Germany, Switzerland, Spain, as well as Caucasia. In the past, einkorn cultivation was much more extensive. This wheat was one of the founder grain crops of Neolithic agriculture in the Near East and a principal component of the early crop assortment in Europe. Since the Bronze Age, its importance seems gradually to have declined, very likely because of the competition from free-threshing wheats. Einkorn is a small plant, rarely more than 70 cm high, with a relatively low yield, but it can survive on poor soils where other wheat types fail. The fine yellow flour is nutritious, but gives bread of poor rising qualities. Thus einkorn has been consumed primarily as porridge or as cooked whole grains. Since Roman times, a considerable part of the yield has been fed to animals (Schiemann 1948; Harlan 1981).

Wild ancestry

The ancestry of the wild progenitor of cultivated einkorn wheat is well established. Cultivated *T. monococcum* is closely related to a group of wild and weedy wheat forms spread over the Near East and adjacent territories and traditionally referred to as wild einkorn or *T. boeoticum* Boiss. emend. Schiem. Both wild and cultivated einkorns are morphologically similar (Fig. 2). Both are diploid ($2n = 14$) and contain identical chromosomes (genomic designation AA). Hybrids between wild *boeoticum* and cultivated *monococcum* are fully fertile and chromosome pairing in their meiosis is normal. The main distinguishing trait between wild einkorn and cultivated einkorn is the mode of seed dispersal. Wild einkorn has brittle ears and the individual spikelets disarticulate at maturity to disperse the seed. In cultivated einkorn the essential adaptation to wild conditions no longer exists. The mature ear remains intact and breaks into individual spikelets only upon pressure (threshing). Survival depends on reaping and sowing by humans. Another diagnostic character indicating domestication is the shape of the grain (van Zeist 1976). In cultivated forms, the kernels tend to be wider compared to the wild forms (Fig. 2).

Because of the close morphological and genetic affinities between wild einkorn and cultivated einkorn and their wide divergence from all other *Triticum* species, most wheat students today regard the wild *boeoticum* wheat not as an independent species but as the wild race of the cultivated crop and place it, as a subspecies, within the crop complex as *T. monococcum* L. subsp. *boeoticum* (Boiss.) A. et D. Löve

Wild einkorn is widely distributed over western Asia and penetrates also into the southern Balkans (Harlan and Zohary 1966; Zohary 1969). Its distribution centre lies in the Near East arc, i.e. northern Syria, southern Turkey, northern Iraq and adjacent Iran, as well as some parts of western Anatolia (Map 1). In these areas, wild einkorn is massively distributed as a component of oak park-forests and steppe-like formations. In addition to occupying such primary habitats, wild einkorn also grows as a weed and colonizer of secondary habitats, such as edges of cultivation and roadsides. Sometimes it also invades fields of cultivated cereals. *Boeoticum* wheats are distributed over a wide ecological range with respect to both soils and climates. Edaphically, wild einkorn shows a definite affinity to basaltic soils, marls, clays, and limestones. Climatically it thrives in the summer-dry foothills of the northern Euphrates basin, as well as on the bitterly cold, elevated plateaux of central and eastern Anatolia (1400–2000 m altitude), with their summer rains, yet it does not stand hot and very arid climates. Farther away from its distribution centre, wild einkorn is restricted mainly to segetal or secondary habitats, i.e. sites which were very probably not available before the opening up of these areas to agricultural activity.

Several major eco-geographical and morphological types can be recognized in wild einkorn. In the north and north-west part of its range, plants with small, one-awned, one-seeded spikelets prevail. These are sometimes named *T. aegilopoides* (Link) Bal. In the summer-dry southern areas, more robust plants with two-grained, two-awned spikelets, sometimes referred to as *T. thaoudar* Reuter, are common. But in central Anatolia, Transcaucasia, and adjacent territories of Iran, a series of intermediate forms, bridging *aegilopoides* and *thaoudar* types, abound. In fact, many Anatolian populations of wild einkorn show a wide range of variation in spikelet morphology and include one-grained individuals (with a single awn), two-grained plants (with two well-developed awns) and various integradations between the two extreme types. The Near East 'arc' also harbours two-grained forms with spread awns, small third awn, reddish kernels, and distinct proteins, which are kept by some wheat students (Miller 1987; Waines and Barnhart 1992) as a separate diploid species: *T. urartu* Tuman. This since crossing experiments indicate that *urartu* forms are intersterile with both cultivated einkorn and typical *boeoticum*.

Archaeological evidence

Einkorn wheat was probably extensively collected from the wild before its introduction into cultivation. Carbonized remains (Fig. 9), agreeing

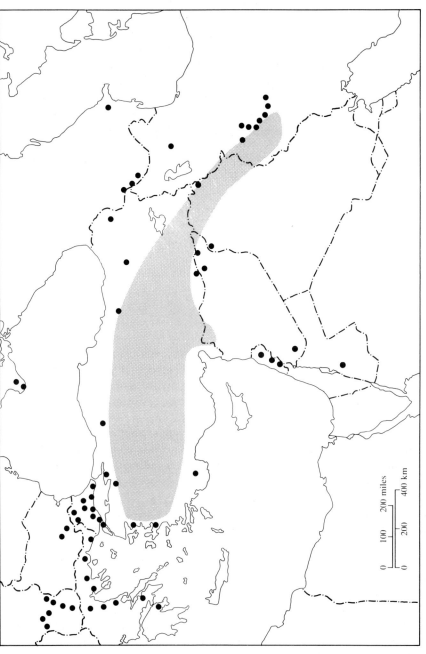

Map 1. Distribution of wild einkorn wheat, *Triticum monococcum* subsp. *boeoticum* (= *T. boeoticum*). The area in which wild einkorn is massively spread is shaded. Dots represent additional sites, outside the main area, harbouring mainly weedy forms. (Based on Zohary 1989.)

Fig. 9. Carbonized grains of wild einkorn wheat, *Triticum monococcum* subsp. *boeoticum.* Final Mesolithic Mureybit, Syria. (van Zeist and Casparie 1968.)

well with present-day wild forms, have been retrieved from 10th and 9th millennia bc Tell Abu Hureyra (Hillman 1975) and from pre-agriculture layers (8th millennium bc) in Mureybit (van Zeist 1970), northern Syria. *Boeoticum*-type einkorn remains continue to appear in some early (7th and 6th millennia bc) Neolithic settlements in the Near East where there are already definite signs of wheat and barley cultivation, i.e. in Tell Abu Hureyra (Hillman 1975), Syria; Çayönü (van Zeist 1972) and Can Hasan (French *et al.* 1972), Turkey; and Ali Kosh and Tepe Sabz (Helbaek 1969), Iran. But in some of these sites, as well as other contemporary Near East settlements, one is faced also with plumper kernels characteristic of cultivated einkorn (Fig. 10), indicating that einkorn belongs to the small group of annual grain plants that founded agriculture in the Near East.

From these localities, which are situated more or less within the present range of distribution of wild einkorn, the cultivated cereal spreads further south to Tell el Aswad (van Zeist and Bakker-Heeres 1985) near Damascus, and to Jericho (Hopf 1983). In both these Early Neolithic sites remains of einkorn are, however, relatively few. This wheat appears to be less frequent than the two other Neolithic founder cereals, namely emmer wheat and barley (Map 2).

Somewhat later, einkorn emerges as one of the principal crops in

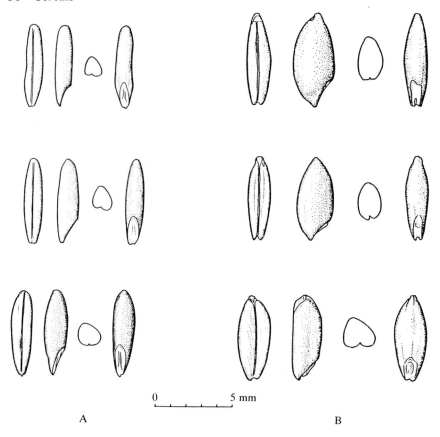

Fig. 10. Comparison between carbonized grains of wild and cultivated einkorn wheats, *Triticum monococcum*. A – Wild einkorn, *T. monococcum* subsp. *boeoticum* from pre-agriculture Mureybit, Syria. (van Zeist and Casparie 1968.) B – Cultivated einkorn, *T. monococcum* subsp. *monococcum*. Early Neolithic Nea Nikomedeia, Greece. (van Zeist and Bottema 1971.)

the now established Neolithic food production in the Near East. It continues to appear together with emmer wheat and barley but shows a definite preference for areas with relatively cool climates. It is rare in warmer places and totally absent from hot regions such as Egypt and Lower Mesopotamia. Einkorn remains continue to be frequent in Chalcolithic, Bronze Age, and Iron Age sites in the Near East. Yet similar to hulled emmer (p.44) it is gradually replaced by the free-threshing wheats.

Einkorn wheat played a major role in the early spread of Neolithic agriculture beyond its Near East nuclear area. Together with emmer wheat, barley, lentil, pea, and flax, this cereal occurs regularly in the Neolithic farming settlements that appear first in Cyprus and Greece and

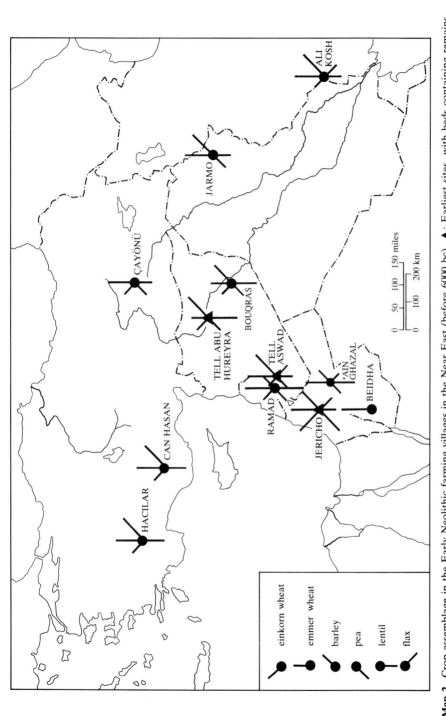

Map 2. Crop assemblage in the Early Neolithic farming villages in the Near East (before 6000 bc). ▲: Earliest sites, with beds containing remains of crops dated to the 8th millennium bc. ●: Somewhat later sites, dated to the 7th millennium bc. A short whisker indicates that the crop is relatively rare and a long whisker that it is relatively common among the excavated plant remains.

subsequently in the Balkan countries, the middle and west Mediterranean basin, central, west, north, and east Europe, and the Caucasus. Carbonized remains of cultivated einkorn are numerous in the 6th millennium bc contexts of Cape Andreas Kastros (van Zeist 1981) and Khirokitia (Waines and Stanley Price 1975–77), Cyprus, and in several early Greek agricultural settlements which developed in the Argolis, Thessaly, and Macedonia during the first half of the 6th millennium bc (Renfrew 1979; Kroll 1981a). But in most sites einkorn finds are less frequent than those of emmer or barley.

Much richer remains of einkorn wheat are available from the sites representing earliest agriculture in Yugoslavia as well as adjacent south Hungary (Starčevo-Körös culture: 5300–4300 bc) and Bulgaria (Karanovo culture: 4800–4600 bc). In these territories, einkorn retains its principal role throughout the Neolithic and Bronze Age (Hopf 1973b; Renfrew 1979). Einkorn is also one of the main cereal crops of the Danubian or Bandkeramik culture (4500–4000 bc), i.e. the first farming settlements in central Europe (Willerding 1980). It usually occurs in a mixture with emmer wheat or sometimes as a pure crop. In some locations both cereals are equally common and occasionally einkorn prevails. This situation persists all over central and northern Europe, also in later Neolithic cultures, although there and then barley becomes more frequent and partly replaces the wheats. Similar dominance of einkorn and emmer is found in eastern Europe but the relative proportions of the two wheats vary considerably according to local conditions and cultures. All in all, the numerous data already available indicate clearly that wheat cultivation in the Neolithic and Bronze Age of central Europe was mainly dependent on hulled types, which do not seem to be replaced here by free-threshing wheats as was the case in the Mediterranean basin and the Near East.

Compared to its principal role in the Balkans and in the central European Danubian (Bandkeramik) culture, einkorn is much scarcer in plant remains obtained from the early Impressed Ware (Cardial) culture in the west Mediterranean basin. Einkorn appears together with emmer and free-threshing wheat in the Cardial settlements in Italy, but rarely as a very common crop. It is scarce in southern France and in the Spanish contexts of the 5th millennium bc and soon afterwards it disappears from these areas altogether. Einkorn also appears in early Neolithic sites in Transcaucasia and Transcaspia (Janushevich 1984; Lisitsina 1984), again as a member of the Near East crop ensemble. But detailed analyses and dating of the material in these areas are still lacking.

Einkorn wheat persisted in many parts of Europe until medieval times. Its cultivation was almost discontinued only in the first half of this century.

Emmer and durum wheats: *Triticum turgidum*

This is a varied aggregate of cultivated tetraploid wheats. All share AABB chromosomes, and all are fully interfertile with one another. According to their response to threshing the cultivated *turgidum* wheats fall into two groups that are frequently recognizable also in archaeological remains:

(i) Hulled emmer wheat, *T. turgidum* L. subsp. *dicoccum* (Schrank) Thell. (traditionally called *T. dicoccum* Schübl.) in which the products of threshing are the individual spikelets (Fig.3). The grains remain invested by the glumes and pales. In emmer, as in einkorn, threshing results in breaking the rachis of the ear in its weakest points below each spikelet. This parallels the disarticulation pattern in wild einkorn. Hulled emmer represents the primitive situation in cultivated *turgidum* wheats.

(ii) The more advanced free-threshing tetraploids (Fig. 4) evolved under domestication from hulled emmer. The common representative of this group is durum wheat, *T. turgidum* conv. *durum* (Desf.) MacKey (syn. *T. durum* Desf.). Less common types are rivet wheat, *T. turgidum* conv. *turgidum* (L.) MacKey (syn. *T. turgidum* L.); Polish wheat, *T. turgidum* conv. *polonicum* (L.) MacKey (syn. *T. polonicum* L.); and *T. turgidum* subsp. *carthlicum* (Nevsky) MacKey (syn. *T. carthlicum* Nevski). Threshing in all these cultivated tetraploid types frees the naked grains. The rachis is usually uniformly tough. Threshing breaks it into irregular fragments.

Hulled emmer *T. turgidum* subsp. *dicoccum* was the principal wheat of Old World agriculture in Neolithic and early Bronze Ages. It was used for food and for brewing high-quality beer. Later it was gradually replaced by more advanced free-threshing tetraploid and hexaploid types. Emmer is still sporadically grown in some parts of Europe and south-west Asia, i.e. the Balkan countries, eastern Czechoslovakia, Anatolia, Iran, Caucasia, and India. In Ethiopia this wheat survives as a common and appreciated cereal.

Durum-type wheats are the main contemporary representatives of the free-threshing AABB tetraploid wheats. Such free-threshing tetraploids (including the small grain forms described by Kislev (1980) as *T. parvicoccum*) had probably appeared already in Neolithic times and gradually gained prominence. Since classical times, free-threshing tetraploid cultivars constituted the main wheat crop in the summer-dry, relatively warm Mediterranean basin. In more continental climates and in temperate areas with summer rains, traditional wheat production depended heavily on another free-threshing wheat: hexaploid *T. aestivum*.

Wild ancestry

Genetic and morphological evidence clearly indicates (for review, see Zohary 1969) that the cultivated tetraploid *turgidum* wheats (both hulled *dicoccum* forms and free-threshing durum varieties) are closely related to a wild wheat native to the Near East and traditionally called *T. dicoccoides* (Körn.) Aarons. (wild emmer wheat). This is an annual, predominantly self-pollinated, tetraploid wheat with characteristic large and brittle ears and big elongated grains (Fig. 3) that show a striking similarity to some cultivated emmer (Fig. 3) and durum (Fig. 4) varieties. It is the only wild stock in the genus *Triticum* that is cross-compatible and fully interfertile with the cultivated *turgidum* wheats. Hybrids between wild *dicoccoides* forms and all cultivated members of *T. turgidum* aggregate show normal chromosome pairing in meiosis, indicating that all these tetraploid wheats contain homologous chromosomes (AABB chromosome sets). The close genetic affinities between wild emmer and the cultivated members of *T. turgidum* are also indicated by spontaneous hybridization that occurs occasionally when these wild and cultivated wheats grow side by side. On the basis of these close genetic and morphological relationships, wheat students now regard *T. dicoccoides* as the wild race of the cultivated tetraploid emmer and durum wheats. They rank wild emmer as the wild subspecies of the *T. turgidum* crop complex (see Tables 3 and 4, pp.24, 29).

In the tetraploid *turgidum* wheats the most conspicuous diagnostic difference between wild and tame is the seed dispersal mechanism (Zohary 1969). Wild *dicoccoides* wheats have brittle ears which shatter upon maturity into individual spikelets. Each spikelet operates as an arrow-like device disseminating the seed by inserting them into the ground. The 'wild-type' rachis disarticulation and the spikelet morphology reflect specialization in seed dissemination which ensures the survival of the wild forms under wild conditions. Under the man-made system of reaping, threshing, and sowing, this adaptation broke down and non-brittle types were automatically selected for. Significantly the shift from a brittle spike (in wild *dicoccoides* wheats) to a non-brittle spike (in cultivated *dicoccum* wheats) is governed by a single major gene. Wild and cultivated forms also differ from one another in kernel morphology (van Zeist 1976). In cultivated *dicoccum* and durum forms the grain tends to be wider and thicker as well as rounder in cross-section compared to the wild *dicoccoides* counterpart (compare grains in Figs 3 and 4). This trait is quite helpful in analysis of archaeological remains.

Wild emmer, *T. turgidum* spp. *dicoccoides* (Körn.) Thell. (syn. *T. dicoccoides* (Körn.) Aaron.), is more restricted in its distribution and more confined in its ecology than wild einkorn. Its range covers Israel, Jordan, Syria, Lebanon, south-eastern Turkey, north Iraq, and west Iran

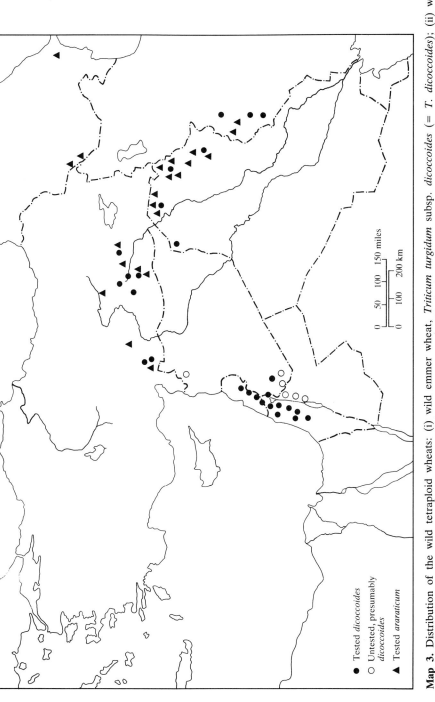

Map 3. Distribution of the wild tetraploid wheats: (i) wild emmer wheat, *Triticum turgidum* subsp. *dicoccoides* (= *T. dicoccoides*); (ii) wild Timopheev's wheat, *T. timopheevi* subsp. *araraticum* (= *T. araraticum*). Black sites represent collections which were tested cytogenetically. (Compiled from Harlan and Zohary 1966; Rao and Smith 1968; Dagan and Zohary 1970; Tanaka and Ishii 1973; Mann 1973; Tanaka *et al.* 1979; and unpublished data of D. Zohary.)

● Tested *dicoccoides*

○ Untested, presumably *dicoccoides*

▲ Tested *araraticum*

(Map 3). It is most widespread in the catchment area of the upper Jordan Valley. It was first found in 1906 in eastern Galilee and on the slopes of Mt. Hermon (Aaronsohn 1910; Schiemann 1956). Wild *dicoccoides* wheats grow as common annual components in the herbaceous cover of the Tabor oak *(Quercus ithaburensis)* park-forest belt and related steppe-like herbaceous plant formations. They are confined to basaltic and hard limestone bedrocks and are completely absent on marls and chalks. In rocky places which have not been severely overgrazed, *dicoccoides* wheat often grow in large stands. With wild barley *Hordeum spontaneum* and wild oat *Avena sterilis*, they form 'fields of wild cereals'.

Further north, wild emmer occurs in the Antilebanon range and in the oak park forest belt of *Quercus brantii* in south-eastern Turkey, north Iraq, and west Iran. Whereas *dicoccoides* wheat occurs alone in the Syrian-Palestine area, it grows sympatrically with a second wild tetraploid wheat, *T. timopheevi* subsp. *araraticum* (see p.54), in the north-eastern part of its distribution area. Cytogenetically, these two wild wheats have distinct genomic and cytoplasmic constitutions and are separated from one another by strong sterility barriers (Maan 1973), but morphologically they are so similar that it is practically impossible to distinguish between them without cytogenetic analysis. Also in the north-east, *dicoccoides* wheat thrives mainly on basaltic and rocky hard limestone slopes. But here it is closely associated not only with *Hordeum spontaneum* but also with wild einkorn, *T. monococcum* subsp. *boeoticum*.

In north Israel and in south Syria, *dicoccoides* wheat shows a multitude of forms and often builds conspicuously polymorphic populations which are easily noticed by their variation in glume hairiness and colour of the spike. Climatically too, wild emmer shows a considerable range of adaptation. It is distributed over a wide altitudinal range. Robust, early maturing types occupy the winter-warm basin around the Sea of Galilee to altitudes as low as 100 m below sea level. More slender, late-blooming forms occur higher up in the Galilee mountains and climb to elevations of 1400 m on the east- and south-facing slopes of Mt. Hermon. In the Zagros Mountains of north Iraq and west Iran, *dicoccoides* wheats occur at altitudes ranging from 700 to 1600 m.

Archaeological evidence

(a) Hulled emmer wheats: Similar to einkorn wheat (p.33) and barley (p.62), emmer wheat was collected from the wild long before its domestication. Brittle, *dicoccoides*-like remains, with relatively narrow grains (Fig. 11) appear in several early Near Eastern settlements. A clear indication of pre-agriculture gathering of wild emmer wheat is available from the south shore of the Sea of Galilee. Here remains of *T. dicoccoides* (as well as wild barley) were recovered from Ohalo

II, a submerged Early Epi-Palaeolithic site dated 17 000 bc (Kislev *et al.* 1992).

The earliest definite signs of cultivated emmer come from Tell Aswad, 25 km south-east of Damascus (van Zeist and Bakker-Heeres 1985). Here numerous plump-seeded, morphologically definable *dicoccum* remains appear in the lowest habitation level Ia (7800–7600 bc), and continue to occur in the two upper phases: Ib (7600–7300 bc) and II (6925–6600 bc). Significantly, no wild *dicoccoides*-like material was retrieved from Tell Aswad; and the climate today in the rain-shadowed Damascus basin is far too dry for wild wheat. Very probably it was arid also 10 000 years ago. As van Zeist and Bakker-Heeres emphasize, the continuous presence of numerous morphologically discernable *dicoccum* remains, the total absence (from the very start) of *dicoccoides* material, and the extreme dryness (less than 200 mm annual rainfall) suggest that emmer wheat was introduced into the Damascus basin already as a domesticated cereal. This not later than 7800 bc. From 7500 bc onward, remains that morphologically conform with *dicoccum* appear also in Tell Abu Hureyra, north-east Syria (Hillman 1975) and in contemporary Pre-Pottery Neolithic B Jericho (Hopf 1983). All these finds indicate that at the beginning of the 8th millennium bc emmer cultivation must have been well under way in the Levant.

From the very beginnings of agriculture in the Near East (second half of the 8th millennium bc and the 7th millennium bc) emmer is the principal cereal of the newly established farming settlements. Although remains of

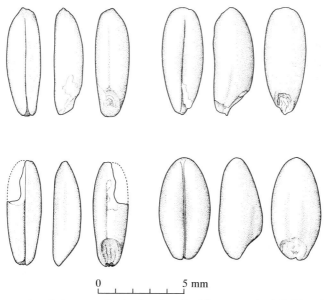

Fig. 11. Comparison between carbonized grains of wild emmer wheat (left) and cultivated emmer wheat (right). Neolithic Çayönü, Turkey. (van Zeist 1972.)

cultivated barley and cultivated einkorn also occur quite regularly in these contexts, emmer prevails quantitatively. In several of the examined sites (Map 2) the remains of emmer include not only carbonized plump seed, but also occasionally well preserved spikelets or 'glume forks' showing the effect of pressure fracturing at the base of the rachis internode. The rough breakage surfaces at the bases indicate that we are confronted with non-brittle ears. This is an unequivocal proof that grain cultivation was already in practice in the Near East arc at that time.

Emmer wheat continues to be the principal grain crop during the later stages of the Neolithic in the Near East, and it was apparently widely grown in this area also in Chalcolithic and Bronze Age times, although in the later periods the hulled wheats (emmer and einkorn) were generally replaced by free-threshing wheats.

Emmer wheat was also the main crop in the spread of the Neolithic agricultural technology from the Near East nuclear area towards the west, the north, the east, and the south. It is the most common constituent of the crop assemblage that started agriculture in the Aegean region (for review, see Renfrew 1979; Kroll 1981a, 1991) and subsequently spread to the Balkan countries (Renfrew 1979, Kroll 1991).

Emmer was the principal cereal of the Danubian (Bandkeramik) farmers that started Neolithic agriculture in central Europe in the 5th millennium bc. At the numerous Bandkeramik sites examined up to date (for review, see Willerding 1980), emmer is usually found side by side or in admixture with einkorn. Yet in most places emmer is the more common wheat. This relation persists all over central and northern Europe during the later Neolithic and the Bronze Age. Thus in central Europe wheat production during the Neolithic and Bronze Age depended heavily on hulled wheats. Naked wheats do not seem to have replaced them so quickly as they did in the Mediterranean basin and in the Near East, though they are present in central Europe since the Neolithic.

Rich finds of emmer have also been discovered at places belonging to the Impressed Ware (Cardial) culture that introduced agriculture in the western Mediterranean basin (Italy, southern France, Spain) during the 5th millennium bc (Hopf 1991a). In Cardial contexts, emmer is generally accompanied by free-threshing wheats and in many localities the latter prevail. Emmer is also found in younger sites of this region, but again, only in minor quantities compared to the more advanced free-threshing wheats. Emmer is, however, the wheat of Neolithic and Bronze Age Egypt (Täckholm 1976; Wetterstrom 1984; Fig. 12).

As to the diffusion of Neolithic agriculture towards the east, there is much less information at hand. Judging by the few data available, emmer wheat was a main element in the (originally Near Eastern) crop assemblage discovered in several Neolithic sites in Caucasia and Transcaucasia, dated into the 5th millennium bc (Janushevich 1984;

Fig. 12. Parched spikelets of cultivated emmer wheat, *Triticum turgidum* subsp. *dicoccum.* (G. Schweinfurth collection, 5th-dynasty Abusir, Egypt.)

Lisitsina 1984). Emmer wheat and barley were the principal grain crops that started Neolithic agriculture in the Indian subcontinent. Their imprints have been discovered in the earliest contexts of Mehragarh, central Pakistan (Jarrige and Meadow 1980; Costantini 1984), i.e. an antecedent site of the Indus valley civilization. These cereals were also principal grain crops of the Harappan civilization in India (Vishnu-Mittre 1977; Kajale 1991).

Finally, it should be pointed out that among the three main Neolithic grain crops (einkorn, emmer, and barley) the wild progenitor of emmer (tetraploid AABB *dicoccoides* wheat) has the most limited distribution; it is confined to the 'arc' of the Near East (compare Map 3 with Maps 1 and 5). Thus the numerous finds of cultivated emmer retrieved from settlements outside the Near East serve as the most convincing evidence for the initiation of Neolithic agriculture in the Near East, and its subsequent spread from this core area.

(b) Free-threshing tetraploid wheats: Free-threshing wheats, identifiable by occasional rachis fragments scattered among the grains (Fig. 8), made their appearance in the Near East soon after the firm establishment of emmer wheat cultivation (van Zeist 1976). They are already present among the plant remains of 7th millennium bc Can Hasan III, Turkey (French *et al.* 1972) and Tell Aswad, Syria (van Zeist and Bakker-Heeres 1985). They also occur in the 6th millennium bc Tepe Sabz, Iran (Helbaek 1969); Choga Mami, Iraq (Helbaek 1969); Çatal Hüyük, Turkey (Helbaek 1964a); and Tell Ramad, Syria (van Zeist and Bakker Heeres 1985).

As argued on p.48, it is impossible to distinguish between remains of tetraploid and hexaploid naked grains. Yet when faced with material retrieved from the early farming villages in the Near East, one can safely assume that they represent 4× *durum*-like forms rather than 6× *aestivum* bread wheats. As argued by Zohary (1969) and by van Zeist (1976), hexaploid bread wheats must have come into existence outside the 'fertile crescent area'; and only after Neolithic wheat agriculture reached the distribution area of *Aegilops squarrosa*. This contact very likely happened in the Caspian belt between 6000 and 5000 bc. Thus the early naked wheats in the Near East core area could not possibly have been 6× *T. aestivum*. They should be attributed to 4× *T. turgidum*.

Naked wheats continue to appear in the Near East in the later Neolithic and in Bronze Age sites, in progressively increasing proportions. They were also retrieved from several 5th millennium bc Greek sites. Naked wheats abound in the Cardial settlements which represent the early stages of agricultural expansion into the west Mediterranean basin (Hopf 1991a). Thus, in the Near East and the Mediterranean basin we are faced rather early with a partial replacement of hulled wheats by free-threshing types. Emmer and einkorn continued to occur; they are occasionally retrieved in large quantities. Yet, towards the Late Bronze Age, the Near East and the Mediterranean basin commonly show a prevalence of naked wheats. The high proportion of naked wheat in the 5th millennium bc Cardial settlements in the western Mediterranean basin is particularly noteworthy. Settlers in these areas apparently preferred free-threshing wheat from the very start. The large quantities of naked wheat unearthed in Bronze Age towns in the Levant are also impressive. In contrast, the wheat of ancient Egypt was hulled emmer and it retained its dominance in the Nile Valley as late as Hellenistic times. It was replaced by naked durum wheats by the Ptolemaic rulers (Wetterstrom 1984).

There is good reason to suppose that the naked wheats retrieved from the late Neolithic and Bronze Age sites in the Near East and the Mediterranean basin (including the small grained forms which are sometimes referred to as *T. parvicoccum*) were also tetraploid, not hexaploid (Harlan 1981). The evidence is circumstantial and it comes from comparative ecology. From classical times until the recent introduction of modern high yielding 6×

wheat varieties, the countries around the Mediterranean Sea were *durum* growing provinces. They produced very little or no bread wheats. *Durum* wheats are basically adapted to Mediterranean environments. In contrast, hexaploid *T. aestivum* thrives in cooler and more continental parts of Europe and western Asia. It is thus reasonable to assume that this traditional geographical distribution of *durum* wheats reflects an even older pattern. In other words, also in the late Neolithic and Bronze Age, the tetraploid wheats should have been the dominant wheat element in the Mediterranean-type environments. By the same rationale one may assume that the remains of naked wheats from Caucasia, Transcaucasia, central and eastern Europe, and central Asia could represent, at least in part, 6× *aestivum* forms. This is if they come from contexts dated later than the 5th millennium bc, i.e. after the formation of hexaploid wheat (see also p.52).

Bread wheat: *Triticum aestivum*

This is the most variable aggregate of cultivated wheats and nowadays economically the most important wheat species. Bread wheat accounts for about 90 per cent of the total world wheat production today and includes numerous and contrasting types. All are hexaploid ($2n = 42$) and interfertile when crossed with one another. All share the AABBDD genomic constitution.

Hexaploid *T. aestivum* is a new wheat species that evolved under cultivation and from the already domesticated tetraploid *T. turgidum* stock. In contrast with the diploid and tetraploid wheats, it does not have a wild hexaploid counterpart. For cytogeneticists and plant evolutionists, bread wheat stands as a classic example of evolution by polyploidy. Genome analysis studies have shown (Kihara 1944; McFadden and Sears 1946) that *T. aestivum* is a hybridization product between a tetraploid *turgidum* wheat (genomic constitution AABB) and a diploid wild grass *Aegilops squarrosa* L. = *Ae. tauschii* Coss. (genomic constitution DD). In other words, hexaploid bread wheats originated by the addition of a third genome to the two genomes already present in *T. turgidum*. Bread wheat has been synthesized in the laboratory by crossing the two putative parents and doubling the chromosomes in the hybrids. Since no AABBDD hexaploid wheat occurs in the wild, this development could have occurred only under cultivation, i.e. by the hybridization of the already domesticated tetraploid wheat with the wild *Ae. squarrosa*. According to their response to threshing hexaploid wheats fall into two groups:

(i) Several types are *hulled*, and today they survive only as relic crops. Prominent among them is spelta wheat, *T. aestivum* subsp. *spelta* (L.) Thell. (traditionally called *T. spelta* L.). An additional type, the macha wheat, *T. aestivum* subsp. *macha* (Dekr. et Men.) MacKey (syn. *T. macha* Dekr. et Men.), endemic to western Georgia, also belongs to this group. In all hulled hexaploid wheats, the products of threshing are the individual spikelets. But while in einkorn and emmer the rachis segment below the spikelet remains attached to the unit, in hulled hexaploid wheats the segment attached is frequently the internode above the base of the spikelet (Fig. 6). In archaeological remains, the upper rachis segment, as well as the characteristic structure of the glumes, serve as diagnostic traits for spelta wheat identification.

(ii) *Free-threshing* bread wheats (Fig.5) are predominant today. Most important and almost universal is bread wheat or common wheat, *T. aestivum* subsp. *vulgare* (Vill.) MacKey (also referred to as *T. aestivum* L., *T. sativum* L., or *T. vulgare* Host). Two other free-threshing hexaploid types are: (a) Club wheat, *T. aestivum* subsp. *compactum* (Host) MacKey (syn. *T. compactum* Host), with compact ears. It is grown today in Afghanistan, as well as in the north-western United States. It has also been found in early European cultures; (b) Indian dwarf wheat, *T. aestivum* subsp. *sphaerococcum* (Perc.) MacKey (syn. *T. sphaerococcum* Perc.), with characteristic small grains is native to India and Pakistan.

Experimental evidence indicates that the first hexaploid wheats were spelta-like. Artificial synthesis of *T. aestivum* (by crossing and fusing tetraploid AABB *T. turgidum* with diploid DD *Aegilops squarrosa*) always results in hulled products (Kerber and Rowland 1974), irrespective of whether the turgidum AABB parent is hulled or naked. This suggests that the free-threshing condition in hexaploid bread wheats was brought about by two events: the appearance of the free-threshing gene q, located on chromosome 5A, and the mutation from Tg to tg in the gene responsible for the tenacious glumes trait on chromosome 2D. All present-day 6× naked wheats examined carry the tg/tg q/q genotype.

Hexaploid wheats have relatively plump grains with blunt tips, particularly the free-threshing forms, while in diploid and tetraploid wheats the kernels are commonly narrower and end with a tapering tip. Yet tetraploid and hexaploid wheats overlap considerably in their seed shape. In archaeological remains, the differences are even more blurred because of swelling and other deformations caused by the charring (Hopf 1955; van Zeist 1976; Harlan 1981). Thus carbonized wheat remains showing obtuse, plump kernels need not necessarily be 6× *T. aestivum*. They can also represent 4× *T. turgidum*. Another diagnostic feature in wheat grains is the shape of the scutellum, i.e. the connecting tissue between the embryo and the endosperm in the grain. It is short, wide, and shows a straight profile in the bread wheats; it is elongated, curved, and shows a curved profile in the durum and einkorn wheats. Yet in this trait one also encounters wide variation and some overlapping. Recently it has been observed (G. Hillman, personal communication) that extreme forms of naked 4× and 6× wheats may be distinguished from one another by the morphology of rachis remains. This might turn out to be a valuable diagnostic tool in the future. All in all, in most archaeological finds one may recognize *free-threshing wheats* as such, but it is impossible to determine whether the grains represent $2n = 28$ chromosomes *durum*-type wheats or $2n = 42$ chromosomes bread wheats. The exceptions are the fortunate rare cases when elements of the spike are retrieved together with the grains.

Wild ancestry

As already stated (p.47), hexaploid bread wheat originated by the addition of the DD chromosome complement of *Aegilops squarrosa* to the cultivated tetraploid AABB *turgidum* wheats. *Ae. squarrosa* (Fig. 13) was never domesticated as such. The examination of its ecology and distribution reveals that this wild grass contributed substantially to the adaptation and the world-wide success of the bread wheats.

The first significant feature about *Ae. squarrosa* is its distribution, which also reflects its climatic requirements (Zohary 1969). This is the easternmost diploid species in the wheat group (*Triticum-Aegilops*). Its centre of distribution does not lie in the Mediterranean Near East but in continental or temperate central Asia. It is widespread and very common in north Iran and adjacent Transcaucasia, Transcaspia, and Afghanistan (Map 4). From this centre, *Ae. squarrosa* spreads westwards to east Syria and eastward to Pakistan. In central Asia it is recorded as far east as Kirghizia and adjacent parts of Khazakstan.

Ae. squarrosa is a very variable species. It is represented by a multitude of forms, from slender types with cylindrical spikes to more robust plants with thick beaded spikes. It also occurs over a wide range of ecological conditions. Like wild einkorn and wild barley it occupies both primary and segetal habitats. In the centre of its distribution this plant is a frequent annual component of open formations. It thrives in areas characterized by continental climatic conditions, from the dry sage-brush steppes of the elevated Iranian and Afghan plateaux to desert margins, as well as in more temperate climates such as the rain-soaked southern coastal plain of the Caspian Sea. At the same time, all over this area *Ae. squarrosa* is a successful colonizer of secondary, man-made habitats and a common weed in cereal fields. Towards the periphery of its distribution, it is almost exclusively a weed in cultivation. Thus, in the case of *Ae. squarrosa* we are faced with a colonizer plant which apparently expanded its distribution over secondary habitats with the opening-up of the land by agriculture.

Fig. 13. Spike of diploid *Aegilops squarrosa* (= *Ae. tauschii*), the donor of the D genome to hexaploid *Triticum aestivum* wheats.

These ecological and distributional facts provide clues to the plausible place of origin of *T. aestivum* and explain some of the bread wheat's ecological specificities. At the start of agriculture, the two contributors that fused to form the hexaploid wheats were evidently geographically separated. Wild emmer was restricted to the Near East arc. *Ae. squarrosa* would not have spread westward from north Iran. Thus contacts between the tetraploid wheats and *Ae. squarrosa* could have been established only

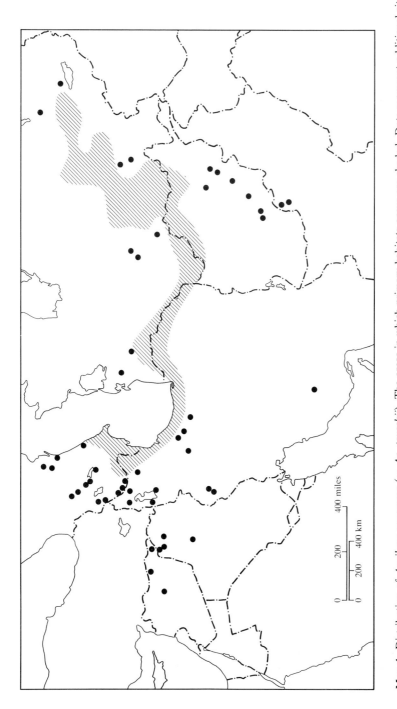

Map 4. Distribution of *Aegilops squarrosa* (= *Ae. tauschii*). The areas in which primary habitats occur are shaded. Dots represent additional sites, mainly of weedy forms. (After Zohary *et al*. 1969.)

after the domestication of emmer and the spread of cultivated tetraploid *T. turgidum* to north Iran and adjacent Transcaucasia. This expansion very probably took place between 6000 and 5000 bc (van Zeist 1976). The most likely area of origin of the hexaploid bread wheat, therefore, is the south-western corner of the Caspian belt. This notion is also supported by isozyme and ribosomal DNA evidence which implicates only one subspecies of *Ae. squarrosa* – confined to the south Caspian Sea belt – as the donor of the D genome to hexaploid bread wheat (Lagudah *et al.* 1993).

The addition of the D genome greatly extended the range of adaptation of wheats (Zohary 1969). Cultivated tetraploid wheats derived from a Near East progenitor were adapted to grow in Mediterranean-type environments with mild winters and warm rainless summers. The incorporation of the *Ae. squarrosa* genome rendered the hexaploid plant more able to withstand continental winters and humid summers. This no doubt facilitated the spread of hexaploid bread wheats over the continental plateaux of Asia and the colder, temperate areas in eastern, central, and northern Europe. It also explains the distribution patterns of wheats in traditional agriculture, i.e. the centring of the 4× *durum* races in the Near East and the Mediterranean basin and the prevalence of 6× *aestivum* forms in temperate Europe and continental west Asia.

Archaeological evidence

(a) Hulled spelta wheat: Spelta has been reported by Janushevich (1984) from 4670 bc contexts of Arukhlo 1, Transcaucasia. The finds are few, but they unmistakably show the tell-tale upper rachis segment attached to the lower spikelet. These data confirm several earlier, poorly documented reports by other Russian workers (Lisitsina 1978) on the occurrence of spelta wheat in some sites in the Kura river plain, Transcaucasia, dating to the 5th millennium bc (and possibly even to the 6th millennium bc). It is therefore clear that in the 5th millennium bc, hexaploid hulled *T. aestivum* was already grown in the Caspian belt. Significantly these early finds come from the area most likely to be the area of formation of hexaploid wheats, i.e. a region in which the cultivated 4× *T. turgidum* could have first come in contact with 2× *Ae. squarrosa*.

Another relatively early spelta wheat find comes from the Neolithic (*c.* 4700 bc) settlement of Sakharova, Moldavia (Janushevich 1984). Here also only a few charred spikelets were detected among numerous remains of emmer wheat. But these clearly display the diagnostic arrangement of the rachis segment of spelta. Sometime later, spelta appears sporadically in east and central European sites. Several well-preserved spikelets, admixed as an 'impurity' with einkorn and emmer remains, were retrieved from Ovčarovo, north-eastern Bulgaria (Gumelnitza culture, *c.* 3750 bc; Janushevich 1978). The well-preserved spikelets leave little doubt that there and then hexaploid spelta wheat was already in use. Similar finds are available from late Neolithic Poland (Schultze-Motel and Kruse 1965) and Germany (Blankenhorn and Hopf 1982). In some Tripolye culture settlements in Moldavia and Bessarabia, spelta wheat was apparently grown as a crop by itself (Janushevich 1978). Spelta wheat remains from Bronze and Iron Age are more numerous and come from all over east, central, and north Europe (Hajnalová 1978) as well as from Greece (Kroll 1983). Spelta was also well known to the Romans (Jasny 1944) and in Europe it survived as a crop until the start of the twentieth century. Very small quantities of spelta are still grown today in south Germany, north Spain, as well as in several other parts of Europe and west Asia.

(b) Free-threshing bread wheat: Genetic analysis indicates (p.48) that naked hexaploid wheats could have evolved rather quickly from their more primitive spelta relatives since the shift was apparently produced by only two mutations. Thus it is plausible to assume that shortly after the formation and establishment of hexaploid spelta the naked derivatives could have appeared as well. In other words, from the 5th millennium bc on, the progressively more common free-threshing wheats encountered in archaeological digs could represent not only tetraploid forms but hexaploid cultivars as well. As already argued previously (p.52), the continental and temperate parts of west Asia and Europe are, ecologically, the territories best suited for the early establishment of the 6× free-threshing bread wheats. It is therefore very likely that from the Bronze Age on, remains of naked wheats from places like Caucasia, central Asia, the central Anatolian plateau, India, and east and central Europe could represent 6× *aestivum* material rather than 4× *turgidum* forms. In the Indus Valley and Baluchistan, the early appearance of plump, very small *sphaerococcum*-type grains in 4000 bc Mehrgarh (Jarrige and Meadow 1980) also indicates hexaploidy, since the *sphaerococcum* mutation is restricted to the hexaploid level. Early literary sources indicate (Ho 1977, p. 449) that wheat was introduced into China only in the 2nd millennium bc. Very likely the introduction was of free-threshing bread wheat.

Timopheev's wheat: *Triticum timopheevi*

This tetraploid species comprises both cultivated and wild forms that are genomically different, and reproductively effectively isolated from the much more common tetraploid *turgidum* stock. Hybrids between *turgidum* and *timopheevi* wheats are sterile, and manifest a considerable amount of chromosomal irregularities in meiosis. On the basis of the available cytogenetic evidence, the genomic constitution assigned to *T. timopheevi* Zhuk. is AAGG. In other words, AAGG *T. timopheevi* shares genome A with AABB *T. turgidum* and AABBDD *T. aestivum*, but does not contain their B genome (Sears 1969; Maan 1973). It also contains a distinct cytoplasm (Maan 1973), which induces male sterility when combined with AABB chromosomes.

Cultivated Timopheev's wheat is hulled, and shows close morphological similarities to cultivated emmer. It is an endemic crop restricted to western Georgia. The wild wheats from which it should have been derived are also well known. The cultigens are fully interfertile and share identical chromosomal constitution with a group of brittle wild wheats scattered over south-eastern Turkey, north Iraq, west Iran and Transcaucasia. These were formerly named *T. araraticum* Jakubz., but are now regarded as the wild race of the cultivated crop, i.e. *T. timopheevi* subsp. *araraticum*. As already mentioned, *araraticum* forms are genetically well isolated from both the wild and the cultivated forms of the tetraploid *turgidum* aggregate. Yet morphologically they show striking similarities to wild emmer, *T. turgidum* subsp. *dicoccoides*. Moreover, in Turkey, Iran, and Iraq both *dicoccoides* and *araraticum* wheats are distributed over the same area (Map 3, p.41) and it is practically impossible to separate them from one another on morphological grounds (Tanaka and Ishii 1973). This can only be done by crossing tests and to a certain extent by examination of seed storage proteins. For these reasons taxonomists dealing with the flora of south-west Asia (e.g. Bor 1968) frequently lump all Kurdish wild tetraploid wheat material together in what they call *T. dicoccoides*, disregarding the fact that they are actually confronted with two reproductively well isolated entities.

From the point of view of domestication and spread of cultivated wheats, the role of the tetraploid *araraticum* AAGG stock is apparently very limited, or even negligible. With the exception of the very localized Georgian *T. timopheevi* cultivars, the chromosomes of the wild *araraticum* stock are absent in the vast ensembles of cultivars and local land races of both tetraploid and hexaploid wheats. It could be argued, of course, that in the early start both wild *dicoccoides* and wild *araraticum* wheats could have been taken into cultivation in south-eastern Turkey, northern Iraq, and western Iran, and that the early non-brittle, hulled wheat remains from these places represent both stocks. Yet the question still remains: if such alleged domesticated AAGG wheats were indeed produced in the Near East, why were they totally replaced by AABB wheats – even among the local land races?

Barley: *Hordeum vulgare*

Cultivated barley, *Hordeum vulgare* L., is one of the main cereals of the belt of Mediterranean agriculture and a founder crop of Old World Neolithic food production. All over this area barley is a universal companion of wheat, but in comparison with the latter it is regarded as an inferior staple and the poor person's bread. But barley withstands drier conditions, poorer soils, and some salinity. Because of these qualities, it has been the principal grain produced in numerous areas and an important element of the human diet. Barley is also the main cereal used for beer fermentation in the Old World. Preparation of this beverage seems to be a very old tradition (Darby *et al.* 1977; Hopf 1976). The crop was, and still is, a most important feed supplement for domestic animals.

Barley is a diploid ($2n = 14$) and predominantly self-pollinated crop. Consequently its variation is structured in true breeding lines. Hundreds of modern varieties and thousands of land races are known. All cultivars have non-brittle ears, i.e. the spikes stay intact after ripening and are harvested and threshed by humans. This is in sharp contrast with wild barleys in which ears are always brittle. Non-brittleness in cultivated barley is governed by a mutation in either one of two tightly linked 'brittle' genes (Bt_1, Bt_2). The brittle wild-type allele in each locus is dominant; the non-brittle alleles are recessive. Many cultivars are homozygous for both recessive mutations. Others carry only one mutation (Takahashi 1955). Non-brittle mutations survive only under domestication, and non-shattering ears serve as a reliable indicator of cultivation.

Barley ears have a unique structure. They contain triplets of spike-lets arranged alternately on the rachis. According to the morphology of the spikelets, barley under domestication can be divided into two principal types:

(i) Two-rowed forms, traditionally called *Hordeum distichum* L., in which only the median spikelet in each triplet is fertile and usually armed with a prominent awn. The two lateral spikelets are reduced, they are borne on longer stalks and are grainless and awnless. Each ear thus contains only two rows of fertile spikelets (Fig. 14B).

(ii) Six-rowed forms, traditionally referred to as *H. hexastichum* L., in which the three spikelets in each triplet bear seed and usually all are awned. Ears in these varieties have therefore six rows of fertile spikelets (Fig. 14C).

The two-rowed condition is primitive; it is found in the wild progenitor of the crop as well as in all other Old World wild *Hordeum* species. Six-rowed types of *Hordeum vulgare* were derived under domestication (Zohary 1969). A single recessive mutation (v) confers fertility to the lateral spikelets and causes the shift from two-rowed to six-rowed ears.

Wild barleys, as well as the majority of the cultivated forms, have hulled grains, i.e. the pales are fused with the kernels and cover them after threshing. In some cultivated varieties this investment is lost. The ripe grains are free and are released when they are ripe by threshing. In

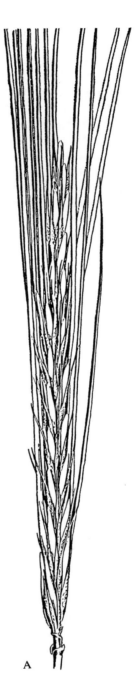

A

Fig. 14. Main types of barley, *Hordeum vulgare*. A – Ear of wild barley, *H. vulgare* subsp. *spontaneum* (= *H. spontaneum*). B – Ear of cultivated two-rowed barley, *H. vulgare* subsp. *distichum*. C – Ear of cultivated six-rowed barley, *H. vulgare* subsp. *vulgare*. D – Grain of hulled barley. E – Grain of naked barley. F – Triplet of six-rowed cultivated barley. Ears 1:1; grains 3:1. (Schiemann 1948.)

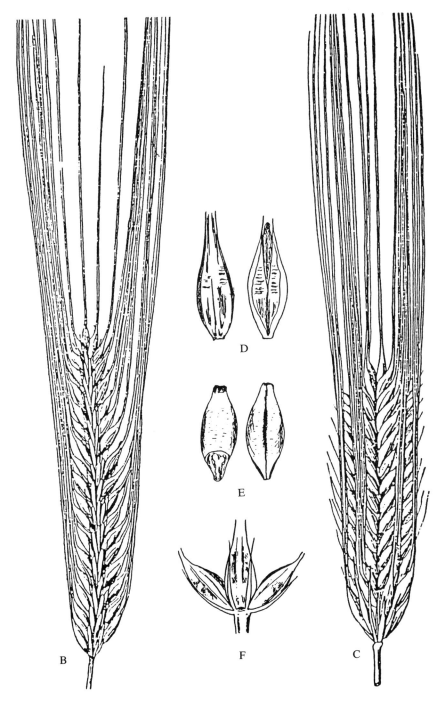

B

C

D

E

F

traditional farming communities naked barleys were frequently favoured for the preparation of food, while hulled forms are preferred for brewing beer and for animal feed. The naked grain trait is controlled by a single recessive gene (n).

Because of the striking differences in ear and kernel morphology taxonomists in the past frequently considered two-rowed barleys and six-rowed barleys as two separate species: *H. distichum* L. and *H. hexastichum* L. = *H. vulgare* L. This nomenclature is still used by some workers. Some barley taxonomists divided the crop even further and called the forms containing fertile lateral spikelets on lax ears *H. tetrastichum* Körn. and those with dense ears *H. hexastichum* L. However, today we know (Zohary 1971; Bothmer *et al.* 1991) that all cultivated barleys contain homologous chromosomes and are fully interfertile. Furthermore, the distinction between the recognized 'species' is brought about by only one of several mutations. Splitting is therefore genetically unjustified and the main cultivated barley types represent races of a single crop species: *H. vulgare* L. (see Table 5).

Carbonized grains are the most common barley remains found in archaeological excavations. Occasionally whole triplets are also retrieved, making it possible to determine whether the cereal was of the two-rowed or six-rowed type. Six-rowed barley forms can also be distinguished from two-rowed barleys by the morphology of the grains: in two-rowed barleys

Table 5.   Taxonomy of the barley crop complex: Species according to traditional classification and their modern ranking on the basis of cytogenetic affinities

Traditional classification	Modern grouping
A section (sect. *Cerealia* Åberg) within the genus *Hordeum* L. containing the following species:	A single species containing both wild and cultivated forms. Collective name: *H. vulgare* L.
1. Wild two-rowed barley *H. spontaneum* C. Koch Brittle, hulled.	1. *H. vulgare* subsp. *spontaneum*
2. Cultivated two-rowed barley *H. distichum* L. Non-brittle, mostly hulled.	2. *H. vulgare* subsp. *distichum* (= *H. vulgare* convar. *distichon*)
3. Cultivated six-rowed barley *H. vulgare* L. (= *H. hexastichum* L.) Non brittle, both hulled and naked forms.	3. *H. vulgare* subsp. *vulgare* (= *H. vulgare* convar. *vulgare*)
4. Brittle six-rowed barley *H. agriocrithon* Åberg.	4. *Agriocrithon* forms are now known to be secondary hybrid derivatives between 1 and 3.

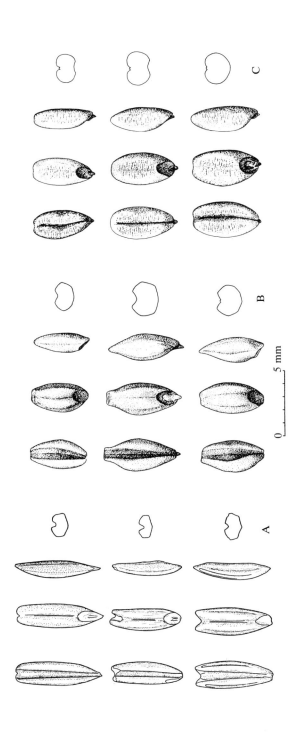

Fig. 15. Comparison between carbonized grains of wild and cultivated barleys. A – Wild barley, *Hordeum vulgare* subsp. *spontaneum* (= *H. spontaneum*) from final Mesolithic Mureybit, Syria. (van Zeist and Casparie 1968.) B – Grains of cultivated barley *H. vulgare* subsp. *vulgare*. C – Grains of naked barley *H. vulgare* var. *nudum*. Medieval Archsum, Germany. (Kroll 1975.)

0 _____ 5 mm

all kernels are straight and symmetrical. In six-rowed barleys the lateral grains are often slightly bent and somewhat asymmetrical. The position and structure of the scars of the spikelets on the rachis internode provide an additional diagnostic trait to identify six-rowed barley in charred remains. Finally, kernels of naked barley (often called var. *coeleste* or var. *nudum*) can be frequently recognized by their shrivelled skin and by the furrow that stays narrow also near the apex (Figs 14 and 15).

Wild ancestry

The wild ancestor of the cultivated barley is well known (Harlan and Zohary 1966; Zohary 1969). The crop shows close affinities to a group of wild and weedy barley forms which are traditionally grouped in *Hordeum spontaneum* C. Koch but which are, in fact, the wild race or subspecies of the cultivated crop. The correct name for this wild type is therefore *H. vulgare* L. subsp. *spontaneum* (C. Koch) Thell. (see Table 5 and Fig. 14). These are annual, brittle, two-rowed, diploid ($2n = 14$), predominantly self-pollinated barley forms and the only wild *Hordeum* stock that is cross-compatible and fully interfertile with the cultivated barley. *Vulgare* × *spontaneum* hybrids show normal chromosome pairing in meiosis. Also morphologically, the similarity between wild *spontaneum* and cultivated two-rowed *distichum* varieties is rather striking. They differ mainly in their modes of seed dispersal. *Spontaneum* ears are brittle and at maturity they disarticulate into individual arrow-like triplets. These are highly specialized devices which ensure the survival of the plant under wild conditions. Under cultivation this specialization broke down and non-brittle mutants were automatically selected for in the man-made system of reaping, threshing and sowing.

The close genetic affinities between the cultivated crop and wild *spontaneum* barleys are indicated also by spontaneous hybridization that occurs sporadically when wild and tame forms grow side by side. Some of such hybridization products, combining brittle ears and fertile lateral spikelets, were in the past erroneously regarded as genuinely wild types and even given a specific rank (*H. agriocrithon* Åberg).

Hordeum vulgare subsp. *spontaneum* is spread over the east-Mediterranean basin and the west-Asiatic countries (Map 5), penetrating as far as Turkmenia, Afghanistan, Ladakh and Tibet. Wild barley occupies both primary habitats and segetal, man-made habitats. Its distribution centre lies in the 'fertile crescent', starting from Israel and Jordan in the south-west, stretching north towards south Turkey and bending south-east towards Iraqi Kurdistan and south-west Iran. In this area, wild *spontaneum* barley is continuously and massively distributed. It constitutes an important annual component of open herbacious formations, and it is particularly common in the summer-dry deciduous oak park-forest belt, east, north,

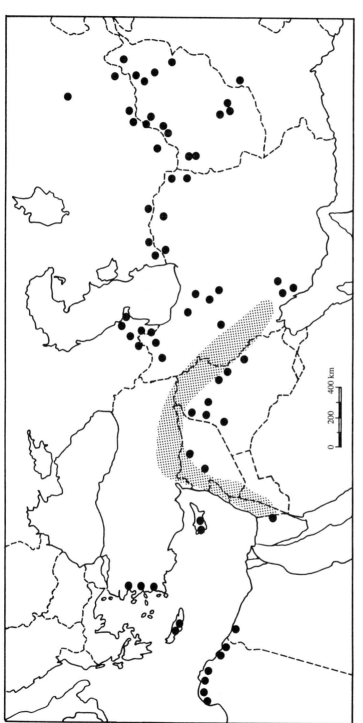

Map 5. Distribution of wild barley *Hordeum vulgare* subsp. *spontaneum* (= *H. spontaneum*). The area in which wild barley is massively spread is shaded. Dots represent additional sites, mainly of weedy forms. Wild barley extends eastwards beyond the boundaries of this map as far as Tibet. (After Harlan and Zohary 1966.)

and west of the Syrian desert and the Euphrates basin, and on the slopes facing the Jordan Rift Valley. From here, *H. vulgare* subsp. *spontaneum* spills over the drier steppes and semi-deserts. In the Near Eastern countries, wild barley also occupies a whole array of secondary habitats, i.e. opened-up Mediterranean *maquis*, abandoned fields, and roadsides. It also infests cereal cultivation and fruit tree plantations (Harlan and Zohary 1966). Further west, in the Aegean region, the Mediterranean shore of Egypt and Cyrenaica and further east in north-east Iran, Central Asia, and Afghanistan, wild *spontaneum* barley is much more sporadic in its distribution; it rarely builds large stands and seems to be completely restricted in most localities to segetal habitats, ruins, or to sites which have been drastically churned by human activity. In general, wild barley does not tolerate extreme cold and it is only occasionally found above 1500 m. It is almost completely absent from the elevated continental plateaux of Turkey and Iran. On the other hand, it is somewhat more drought resistant than the wild wheats and penetrates relatively deeply into the warm steppes and deserts.

Archaeological evidence

Barley first appears in several pre-agriculture or incipient sites in the Near East (17 000–8000 bc). The remains are of brittle, two-rowed forms, morphologically identical with present-day wild *spontaneum* barley, and apparently collected from the wild (Fig. 15). The earliest record of such wild barley harvest comes from Ohalo II, a submerged Early Epi-Palaeolithic site on the south shore of the Sea of Galilee. Here remains of *H. spontaneum* were retrieved (together with wild emmer wheat) and dated to *c.* 17 000 bc (Kislev *et al.* 1992). Other early signs of *H. spontaneum* come from 9000 bc Tell Abu Hureyra (Hillman 1975) and from 8000–7500 bc Mureybit, north Syria (van Zeist 1970) and 7800–7300 bc, Tell Aswad east of Damascus (van Zeist and Bakker-Heeres 1985). Similar finds are available from the 7500–6750 bc. Bus Mordeh phase of Ali Kosh, Iran (Helbaek 1969), the earliest layers (7000 bc) of Çayönü, Turkey (van Zeist 1972) and from 6700 bc pre-ceramic Beidha, Jordan (Helbaek 1966a). At the last three sites, brittle *spontaneum* type barley was found in contexts already showing definite signs of wheat cultivation.

Unmistakable remains of non-brittle barley, i.e. forms that could survive only under cultivation, appear at the 8th millennium bc and in the 7th millennium bc. The oldest indication comes from 7750 bc Pre-Pottery Neolithic A Netiv Hagdud north of Jericho (Kislev *et al.* 1986). Here numerous remains of non-brittle barley, mixed with brittle *spontaneum*-type, were uncovered. Although their proportion in the barley remains is only 12 per cent they might represent cultivation. Other early finds of non-brittle barley came from Aceramic Neolithic (7500 bc onward) Tell

Abu Hureyra (Hillman 1975), phase II (6900–6600 bc) in Tell Aswad (van Zeist and Bakker-Heeres, 1985) and from 7000–6500 bc pre-pottery Jarmo, Iraq. In the latter site, Hans Helbaek (1959b) was the first to show two-rowed barley remains still closely resembling wild *spontaneum* but also displaying a non-brittle rachis. Similar finds were reported by Hopf (1983) in pre-pottery Jericho. Indicative clues come also from Ali Kosh (Helbaek 1969), where the brittle *spontaneum*-like material characterized the lower layers, and in the upper strata it was replaced by non-brittle, broad-seeded, barley forms.

Cultivated barley continues to be a principal grain crop in the Near East throughout the Neolithic period. Its remains have been recovered, side-by-side with wheats, in practically every Neolithic site in which rich plant remains were retrieved. Very soon we are faced also with more advanced forms, i.e. six-rowed hulled as well as naked cultivars of barley.

Six-rowed barley starts to appear already very early. Remains of this barley appear in the aceramic Neolithic beds (7500 bc onward) of Tell Abu Hureyra (Hillman 1975). Information on its emergence also comes from Ali Kosh (Helbaek 1969). In the earliest phase (7th millennium bc) of this site only two-rowed barley occurs but, from about 6000 bc on, sporadic six-rowed elements, as well as naked kernels, appear among the otherwise two-rowed material. Similar remains of six-rowed barley were found in 5800–5600 bc Tell-es-Sawwan (Helbaek 1964b). In Anatolia, six-rowed barley was already firmly established in the 6th millennium bc. In Çatal Hüyük and Hacilar it is represented by both hulled and naked varieties (Helbaek 1964a, 1970). Later, in the 4th millennium bc, hulled six-rowed barley establishes itself as the main cereal of the Mesopotamian basin. Hulled barleys are common in the Near East also in Chalcolithic and Bronze Age contexts. In these periods they show a tendency to outnumber the wheats. Being less sensitive to changes in climate and soils, they adapted more easily to extreme conditions and very probably replaced wheats on depleted soils or in irrigated areas suffering from salinization.

Barley was one of the principal crops of the subsequent spread of agriculture from the Near East, first to the Aegean region and subsequently to the Balkan countries, central Europe, the west Mediterranean basin, as well as Egypt, Transcaucasia, and the Indian subcontinent.

During the 6th and 5th millennia bc, barley emerges in Greece as a constant companion of emmer and einkorn wheats. It is represented by both two-rowed and six-rowed forms and also by naked varieties (practically all Neolithic naked barleys are six-rowed). In Chalcolithic and Bronze Age Greece, barley is again gaining in importance and frequently becomes the prevailing cereal. In Neolithic cultures in the Balkans and central Europe one encounters, from the early beginnings, mainly six-rowed forms and mostly naked grains. Barley is relatively rare in the Danubian (Bandkeramik) culture. In some sites naked grains are as common as hulled ones. But also in central Europe (as well as northern Europe) its importance is increasing in later Neolithic times and especially in the Bronze Age.

In the Cardial and later Neolithic cultures of the west-Mediterranean basin (5th through 3rd millennia bc), barley remains are plentiful and they consist mainly of hulled and naked six-rowed forms (Hopf 1991a). It remains a close associate of free-threshing wheats in these territories throughout the Bronze Age.

Barley is the main companion of emmer wheat in the Neolithic settlement of the Nile Valley in the 5th millennium bc (Darby *et al.* 1977; Wetterstrom 1984) and also in this area it maintained its important role in food production through Neolithic and Bronze Age times. Barley also seems to be a main element in the diffusion of the Near East agriculture towards the east. It is present in the crop assemblage in several sites in Caucasia and Transcaucasia dating back to the 5th millennium bc (Lisitsina 1984). Barley is also a major crop in 6th and 5th millennia bc Mehrgarh, i.e. the oldest known agriculture in the Indian subcontinent (Jarrige and Meadow 1980; Costantini 1984). Later, together with wheat, barley appears to be an important constituent in the Harappan civilization crop assemblage (Vishnu-Mittre 1977; Kajale 1991). Similar to wheat (p. 53), barley seems to have reached China rather late. Unquestionable documentation of its use in this part of the world comes only from the second half of the 2nd millenium bc (Ho 1977, p. 449).

In conclusion, the archaeological finds show barley as a founder crop of the Near Eastern Neolithic agriculture and as a close companion of emmer and einkorn wheats. It is also clear that wild *spontaneum* barley is the ancestral stock from which cultivated barley was derived. This wild barley is a common annual plant in the Near East arc and the area of its distribution coincides well with the sites of the earliest finds of barley cultivation. This coincidence indicates quite conclusively that the Near East is the place of origin of cultivated barley. The archaeological remains make it also possible to trace the main developments of barley under domestication: first the fixation of non-brittle mutations and subsequently the emergence of six-rowed hulled, and naked types. The principal role of barley in food production in the Old World in Neolithic and Bronze Age times is now well documented.

Rye: *Secale cereale*

Rye, *Secale cereale* L., is a characteristic grain crop of the temperate areas of the Old World. It is particularly appreciated in northern and eastern Europe because of its winter hardiness, resistance to drought, and its ability to grow on acid, sandy soils. Thus it succeeds under conditions in which wheat frequently fails (Evans 1976). Rye grains contain appreciable amounts of proteins and can be baked into dark-coloured 'rye-bread', yet the baking quality of the flour is inferior to that of wheat. Much of the present world production of rye is consumed in the form of bread and the grains are also used as a high energy animal feed and for the preparation of rye whiskey. The green plants are also commonly used for fodder. In contrast to most grain crops which are self-pollinating, rye is a cross-pollinated cereal. Yields depend among other factors on effective wind pollination.

Wild ancestry

Cultivated rye belongs to a small genus of grasses, *Secale* L., the distribution of which is centred in south-west Asia. Recent classifications (Sencer and Hawkes 1980) recognize three biological species in this genus. All are diploids with $2n = 14$ chromosomes.

Closest to the crop is a variable aggregate of annual weedy and wild ryes. They are fully interfertile and chromosomally homologous with the cultivated rye varieties and are included in the crop complex, *S. cereale* L.

A variable group of perennial ryes, placed in *S. montanum* Guss., stands not very far from the crop; but differs from it by two chromosomal translocations. Hybrids between annual *cereale* forms and perennial *montanum* types are semi-sterile and manifest a characteristic translocation hexavalent in meiosis (Stutz 1972). In spite of this chromosomal divergence, reproductively *S. cereale* and *S. montanum* are not totally isolated from one another (see p.67).

Finally, *Secale* contains a third fully diverged species, *S. sylvestre* Host. This small, self-pollinated wild annual grows in the Black Sea and Caspian Sea basins. Morphologically and genetically, *S. sylvestre* is distinct from the crop and it seems to have played no part in the evolution of *S. cereale* under domestication.

The following paragraphs summarize the available information on *S. cereale* and *S. montanum* aggregates.

Secale cereale: As already mentioned, the crop complex contains a variable assortment of cultivars, weedy races, and wild types. All are annual and most of them are cross-pollinated. Hybrids between the various forms show normal formation of seven bivalents in meiosis and are fully fertile. In the past, rye taxonomists (Roshevitz 1947) split this complex into several species. The cultivated varieties were grouped in *S. cereale* L., while the variable weeds and wild types were placed in *S. segetale* (Zhuk.) Roshev.,

S. afghanicum (Vav.) Roshev., *S. dighoricum* (Vav.) Roshev., *S ancestrale* Zhuk., and *S. vavilovii* Grossh. Yet the results of comprehensive genetic tests (Nürnberg-Krüger 1960a, b; Khush 1963a; Stutz 1972; Vences *et al.* 1987) indicated that such splitting is unjustified. The types described are now regarded as main races or subspecies of the crop complex (Sencer and Hawkes 1980). There is, however, one exception: a *vavilovii*-like collection from Hamadan, Iran, which is chromosomally different from all other forms (Kranz 1961; Khush 1963b; Sencer and Hawkes 1980) and may represent an additional, not fully explored annual *Secale* species. The variable *cereale* complex can be roughly divided into the following principal races:

(i) *Cultivated varieties*: These are non-shattering plants with character-istic large and plump grains (Fig. 16C).

(ii) *Non-shattering weeds*: This is a variable aggregate of obligatory weeds infesting wheat cultivation in Turkey, adjacent areas in Syria, Iraq, and Iran, as well as the Balkan countries, Caucasia, and Transcaucasia. These have frequently been referred to as *S. segetale*. The mature ears of these weeds do not shatter and they also resemble wheat in grain size and grain weight. Consequently they are harvested and threshed together with the wheat. Since traditional winnowing cannot separate grains of *segetale* ryes from those of wheat, rye seed is included in the harvest and planted with the wheat in the subsequent year. Farmers, particularly those in the elevated plateau of Anatolia, tolerate some rye weed mixture in their wheat crop, and for a good reason: in bad years with extreme cold and dry weather the rye weeds survive when wheat does not – ensuring the peasants of what is sometimes referred to as the 'wheat of Allah'.

(iii) *Semi-shattering weeds*: These weedy ryes are common in north-east Iran, Afghanistan, and adjacent Central Asian republics. They also infest grain cultivation. The rachis is semi-brittle; the upper part of the mature ear shatters spontaneously while the lower part stays intact and is reaped together with the wheat crop. Different populations vary in the intensity of ear shattering and not infrequently two-thirds of the seed produced undergoes 'wild type' dissemination and stays in the cultivated field. Semishattering weedy ryes have frequently been referred to as *S. afghanicum* (Vav.) Roshev.

(iv) *Fully shattering weeds and wild types*: In these forms the rachis is fully fragile and the mature ear disarticulates spontaneously into individual spikelets. The kernels are narrow and fully covered by the glumes. The individual wedge-like spikelets serve as seed dispersal devices. The first brittle rye was discovered near Aydin, western Turkey, by the Russian botanist P.M. Zhukovsky, and named by him *S. ancestrale* (Fig. 16B). But since this rye infests fig plantations and vineyards, it is obviously a weed and not a genuinely wild type. Truly wild, brittle annual ryes that are chromosomally homologous with the cultivated crop occur, however, in eastern Turkey and adjacent Armenia (Map 6). Compared to the robust *ancestrale* weeds, they are low-growing and are commonly referred to as *S. vavilovii* (Stutz 1972; Sencer and Hawkes 1980). They are locally common on the lower slopes of Mt. Ararat where they thrive on basalt slopes and on volcanic ashes. Similar low-growing fully brittle annual ryes were recently found by one of us (D. Zohary) growing in the basaltic escarpments of Mt. Karacadagh, west of Diyarbakir, south-east Turkey.

Secale montanum comprises a variable group of perennial cross-pollinated ryes spread over the elevated plateaux and mountains of the Near East (especially Turkey) and Caucasia and distributed more sporadically in several mountain ranges in Transcaspia and in the central and western parts of the Mediterranean basin (Sencer and Hawkes 1980). Perennial *montanum* ryes have a tufted growth habit and appressed shattering ears (Fig. 16A). Russian botanists have split the perennial ryes, too, into several species: *S. montanum* Guss. in the Mediterranean basin, *S. dalmaticum* Vis. in Dalmatia, *S. anatolicum* Boiss. and *S. ciliatoglume* (Boiss.) Grossh. in Anatolia, and *S. kupriyanowii* Grossh. in Iran and Transcaucasia. Yet all these taxa intergrade with one another, are fully interfertile and seem to represent only major eco-geographic races. Consequently, they are regarded in recent treatments of *Secale* (Sencer and Hawkes 1980) as only infra-specific units and are lumped together in *S. montanum*. Closely related to the south-west Asiatic and Mediterranean perennial ryes is the taxon *S. africanum* Stapf., distributed over a small area in South Africa. Very probably it represents a geologically recent arrival of rye to this widely disjunct territory.

Over considerable areas on the elevated plateaux of Anatolia, *S. montanum* grows side by side with *S. cereale*, the first as a common grass in the non-arable steppe habitats and the second as a weed in adjacent cultivated fields. Spontaneous hybridization between these two cross-pollinated grasses was found to be rather frequent in such contact places, particularly at the edges of cultivation (Stutz 1972; D. Zohary, unpublished data). Introgressive hybridization variation patterns often characterize both species in these areas. Thus, in their main geographic centre, *S. montanum* and *S. cereale* are not reproductively fully isolated from one another and they still exchange genes.

A

Fig. 16. Ears and grains of wild and cultivated ryes. A – Perennial *Secale montanum*. B – Brittle *S. cereale* subsp. *ancestrale*. C – Cultivated rye, *S. cereale* subsp. *cereale*, variety Petkus. Ears 1:1, grains 3:1. (Schiemann 1948.)

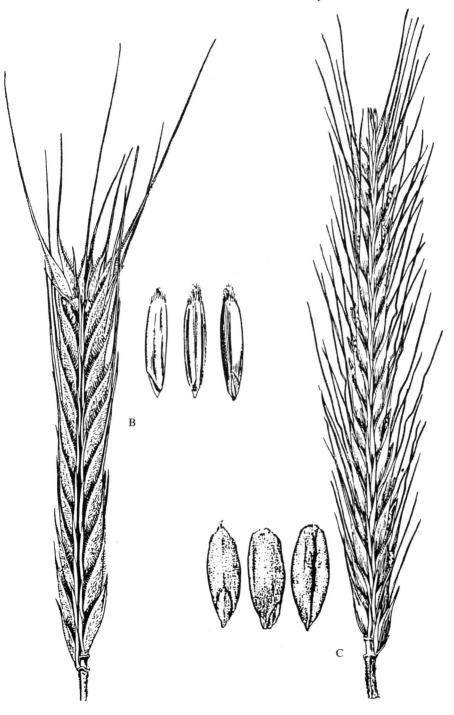

B

C

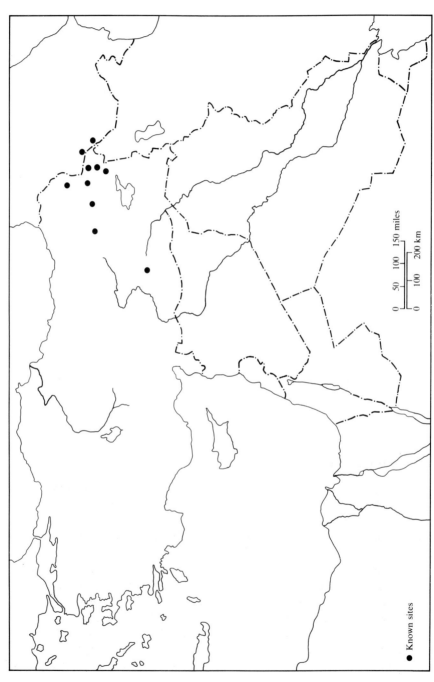

Map 6. Distribution of brittle wild rye, *Secale cereale* subsp. *vavilovii* (= *S. vavilovii*). (Compiled from Sencer and Hawkes 1980; Stutz 1972; and unpublished data of D. Zohary.)

● Known sites

0   50   100   150 miles

0   100   200 km

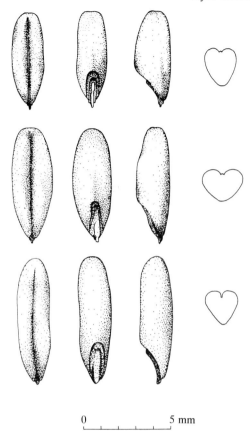

0 _____ 5 mm

Fig. 17. Carbonized grains of cultivated rye, *Secale cereale*. Medieval Archsum, Germany. (Kroll 1975.)

Archaeological evidence

Only very few remains of *Secale* have been discovered in the Neolithic and Bronze Age settlements in the Near East. This is in spite of the fact that Turkey and Iran harbour today a wealth of wild and weedy forms of *S. cereale*, and that the charred grains of this cereal can be easily identified (Fig.17).

The earliest remains come from Epi-Palaeolithic Tell Abu Hureyra, north Syria (Hillman 1975). They are brittle and apparently represent *S. montanum* material collected from the wild. Charred remains of rachises and grains of rye were recovered from aceramic Neolithic Can Hasan III, Turkey, by screening bulk deposits by flotation (Hillman 1978). The kernels conform well with grains of primitive cultivars or weedy *S. cereale* found in Anatolia today, and the rachises are non-brittle. These finds indicate that in Anatolia rye entered cultivation already in early Neolithic times either as a man-dependent weed or as a crop. Yet no additional rye remains have been discovered in other Neolithic Near East sites. The next record comes from Bronze Age levels of Alaca Hüyük in north-central Anatolia (Hillman 1978). Here a pure hoard of carbonized large grains of *S. cereale* was discovered indicating that there and then rye was grown as a crop in its own right.

Outside the Near East, information on rye is again very fragmentary. The earliest reports are from several late Neolithic Funnel Beaker (TRB) culture sites in north, central, and southern Poland (Wasylikowa *et al.* 1991), and from late Danubian Vösendorf, Austria (Werneck 1961). But only very few grains were found in the Polish sites compared to the abundance of wheats or barleys. Furthermore, most records are imprints – a fact that makes their identification less reliable. In Vösendorf, only a few broken carbonized kernels were discovered. Their identification is again questionable since it is based solely on measurements of cross-sections. It is therefore uncertain whether these European finds represent a crop, rye weed, or rye at all (Hillman 1978).

The first unequivocal evidence on rye cultivation in Europe comes from several Bronze Age (1800–1500 bc) settlements in Czechoslovakia where Tempír (1966, 1968) discovered rich remains of carbonized grains. Bronze Age records of rye are available also from Moldavia and the Ukraine (Wasylikowa *et al.* 1991). They all represent rye grains contaminating other cereals. Some what later, rye appears in Iron Age settlements in Germany (Hopf 1982, table 1), Denmark (Helbaek, 1954), Poland (Willerding 1970) and Crimea (Janushevich 1978) – usually admixed with barley or wheat. Also in Hasanlu, Iran (Tosi 1975), *S. cereale* appears at the end of the 2nd millennium bc and seems to have been a staple crop throughout the Iron Age. Rye was part of Roman grain agriculture and was grown in the cooler northern provinces. Carbonized rye grains have been retrieved from several Roman frontier sites along the Rhine and the Danube (Hillman 1978), as well as from the British Isles (Jessen and Helbaek 1944).

In summary, the available evidence from the living plants points to the annual, brittle *vavilovii* ryes as the wild ancestor of the cultivated crop; and to eastern Turkey and adjacent Armenia as the most probable place of origin. The archaeological record supports this notion. The earliest sign of rye, associated with agriculture, comes from central Anatolia. Compared to wheat and barley, remains of rye in archaeological digs are discouragingly few. Additional finds are necessary in order firmly to establish the place of origin and the subsequent pattern of spread of this crop.

The available evidence also suggests that rye evolved first as a tolerated weed, and was only later picked up as a crop. The data also indicate that variation build-up in this cereal and the impressive evolvement of weedy ryes could have been considerably enhanced by introgressive hybridization with perennial *montanum* ryes, i.e. the common wild rye element distributed over the elevated, continental parts of Anatolia and adjacent areas in the Near East.

Common oat: *Avena sativa*

Common oat, *Avena sativa* L., constitutes a major cereal crop in traditional Old World agriculture and a close companion of wheats and barley. The crop succeeds well in moist climates of temperate latitudes. In such environments, particularly in north-west Europe, oat frequently thrives better than wheat and is cultivated as a principal grain crop (Holden 1976). The nutritive value of oat is high. The grains contain about 15–16 per cent protein and 8 per cent fat. They serve as a staple in human diet and as a high energy supplement for farm animals.

Three cytogenetically distinct stocks of *Avena* L. occur under cultivation. Each had an independent origin (Zohary 1971). Yet only one species, the hexaploid ($2n = 42$) common oat, *A. sativa*, established itself as a principal cereal. Two other oats are minor crops of negligible significance:

(i) Cultivated diploid ($2n = 14$) *A. strigosa* Schreb. (including the diploid cultivars traditionally named *A. brevis* Roth), is grown as a fodder plant in some parts of Western Europe. The crop originated from wild forms of *A. strigosa*. The latter are widely spread over the western parts of the Mediterranean basin.

(ii) Tetraploid ($2n = 28$) *A. abyssinica* Hochst. is a half-weed, half-crop confined to the highlands of Ethiopia. It is derived from *A. barbata* Pott., a wild and weedy tetraploid oat widely distributed in the Near East and in the Mediterranean basin.

The common oat, *A. sativa*, comprises numerous and contrasting cultivated varieties. All are interfertile with one another and share the same hexaploid genomic constitution. All are also characterized by non-shattering panicles. According to their response to threshing they were traditionally placed in three 'species', which are in fact only morphological types of the same crop species. Most cultivars are hulled, i.e. their grains remain invested by the pales. In some of these hulled forms (*A. sativa* in the narrow sense) threshing results in pressure breakage of the spikelet's rachilla at the base of the upper floret. The rachilla segment remains attached to the lower floret. In other cultivars (traditionally named *A. byzantina* C. Koch), the rachilla segment breaks at its base and remains attached to the upper floret. The basal floret shows an abscission scar. In several hexaploid cultigens (frequently called *A. nuda* L.) the grains are free and threshing releases the naked kernels. Nakedness is a derived trait and occurs only under domestication.

Wild ancestry

Avena L. is a relatively small Mediterranean genus comprising some 12 annual species. The majority of the oat species have a diploid ($2n = 14$) chromosome number; three are tetraploids ($2n = 28$) and one is hexaploid ($2n = 42$).

The 6× cultivated *sativa* oats show tight genetic affinities and close morphological similarities to a group of wild and weedy oats, which are widely distributed over the Mediterranean basin and traditionally called *A. sterilis* L. (Fig. 18) and *A. fatua* L. These wild types are also hexaploid ($2n = 42$) and contain chromosomes homologous to those present in the cultivars. Moreover, all wild and cultivated 6× oats are interfertile and occasionally cross in nature. Because of these close affinities, *sterilis* and *fatua* oats are now recognized as the wild races of the cultivated crop and are placed within the *A. sativa* crop complex (Malzew 1930; Ladizinky and Zohary 1971; De Wet 1981). The *sativa* complex is, however, chromosomally distinct and reproductively isolated from all other *Avena* species. It is also the only hexaploid member in this genus.

Just as in wheats and barley, domestication brought about a breakdown of the original way of seed dispersal. Wild and weedy oats produce highly specialized drill-type diaspores, with characteristic kinked awns, in order to disseminate their seed and to insert them into the ground. They shed their seed immediately after maturation. As already noted, cultivated varieties have non-shedding panicles. Reduction of the awns (compare Fig. 18B with Fig. 19A) and the evolution of relatively compact panicles are additional conspicuous developments under domestication.

Two distinct modes of seed dispersal operate in the wild members of the *A. sativa* complex and serve as a diagnostic trait for their classification. Most widespread are oats in which the spikelet shows a single disarticulation point at the base of the lower floret. Consequently the whole spikelet (minus the glumes), with two to three invested kernels, constitute a drill-type dispersal unit. This fruiting type was traditionally named *A. sterilis* L. In other forms, conventionally placed under *A. fatua* L., the rachilla of the spikelet disarticulates at each node; the florets, each with a single seed, disperse individually.

Sterilis-type oats massively colonize the Mediterranean basin from the Atlantic coast of Morocco and Portugal in the west to the Zagros

Fig. 18. Wild oat, *Avena sativa* subsp. *sterilis* (= *A. sterilis*). A – Fruiting panicle. B – Seed dispersal unit (spikelet without glumes). (Malzew 1930, plates 86, 88.)

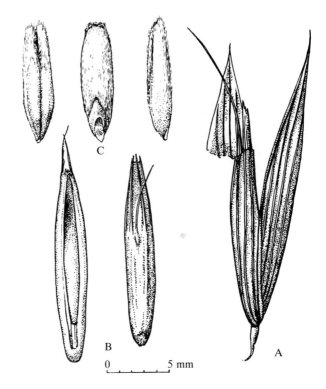

Fig. 19. Cultivated oat *Avena sativa*. A – Non-shattering spikelet. B – Lower and upper florets. C – Grain.

Mountains in the east. In the Near East they frequently grow together with wild wheats and barley. In addition to massive occupation of primary habitats *sterilis* oats aggressively colonize abandoned cultivated ground all over the Mediterranean region; and they grow as a noxious weed in wheat and barley cultivation, orchards, and roadsides. *Sterilis* oats occur in masses in areas with typical Mediterranean climate and are also morphologically very variable. Most conspicuous is the variation in the size of the spikelet, its hairiness, and colour. Forms with smaller spikelets are sometimes referred to as *A. ludoviciana* Dur. and those with larger spikelets and three to four fertile florets are called *A. macrocarpa* Mönch.

Fatua oats are distinctively weedy. They are widely distributed over the whole belt of Old World agriculture where they infest cereal fields and grow at the edges of cultivation. Only rarely do they occupy primary habitats. *Fatua* forms also thrive in colder, more continental climates. In elevated mountain cultivation and on the northern fringes of west Asiatic grain agriculture *fatua* plants sometimes replace *sterilis* oats entirely.

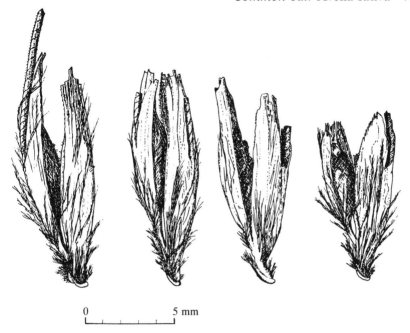

0 _____ 5 mm

Fig. 20. Carbonized spikelets of wild oat, *Avena sativa* subsp. *sterilis* (= *A. ludoviciana*), from Auvernier, Switzerland. (Villaret-von Rochow 1971.)

Archaeological evidence

There is no sign of oat cultivation in Neolithic or Bronze Age sites in the Near East and the Mediterranean basin. This is in spite of the fact that wild *sterilis* oats are massively distributed over these territories and even grow together with wild wheats and wild barley. The few oat remains retrieved from Neolithic Near East and European sites seem to represent only shattering wild or weedy *sterilis* or *fatua* forms (Fig. 20). Definite indications of domestication, i.e. remains of non-shattering *sativa* or *byzantina* plants with their characteristic plump seed (Fig. 21) appear first in Europe – but only in the 2nd and 1st millennia bc contexts (Willerding 1970, pp.345–6; Villaret-von Rochow 1971). Among the early remains of cultivated oats are finds from 1st millennium bc Czechoslovakia (Tempír 1966).

In conclusion, the data available from the archaeological digs, as well as the evidence obtained on the ecology and distribution of the wild relatives, agree well with the notion that cultivated *A. sativa* should be regarded as a secondary crop. Very probably, oat started its evolution under domestication not as a crop but by evolving weedy types that infested wheat and barley cultivation. Only much later were such weeds picked

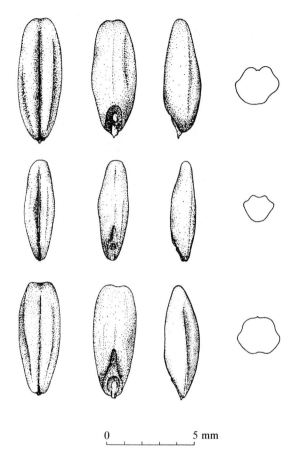

0 _____ 5 mm

Fig. 21. Carbonized grains of cultivated oat, *Avena sativa*. Medieval Archsum, Germany. (Kroll 1975.)

up and planted intentionally. In the temperate periphery of Old World agriculture, oats not only supplemented wheat and barley, but also evolved into a principal grain crop.

Broomcorn millet: *Panicum miliaceum*

Broomcorn millet or common millet, *Panicum miliaceum* L. (Fig.22), ranks among the hardiest cereals. It is a warm season plant which stands up well to intense heat, poor soils, and severe droughts, completing its life cycle in a very short time (60–90 days) and succeeding in areas with short rainy seasons. This is the true millet of classic times (the Romans' *milium* and the Hebrews' *dokhan*). Today, *P. miliaceum* is grown mainly in eastern and central Asia, in India and (more locally) in the Middle East. The dehusked grains are boiled and cooked like rice, or ground for the preparation of porridge. They are quite rich (10–11 per cent) in proteins. The seeds are also used as bird-feed. Broomcorn millet has a tetraploid ($2n = 36$) chromosome complement. It is mainly self-pollinated.

The wild ancestor of the cultivated *P. miliaceum* is not yet satisfactorily identified. Weedy forms of this millet are widespread in central Asia,

Fig. 22. Broomcorn or Common millet, *Panicum miliaceum*. A – Fruiting panicle; B – Ripe spikelets. (Hegi 1935, p. 263.)

from the Aralo Caspian basin in the west to Sinkiang and Mongolia in the East. They are referred to as *P. miliaceum* subsp. *ruderale* (Kitag.) Tzvelev = *P. spontaneum* Lyssov ex Zhuk. Recently these weeds, with their characteristic shattering panicles, spread to central Europe and North America (Scholz 1983). Very probably the vast semi-dry areas in central Asia harbour not only weedy, but also genuinely wild *miliaceum* forms.

No archaeobotanical documentation on *P. miliaceum* is available from central Asia, i.e. the area in which this millet grows wild. However, remains of millet have been reported from several Eneolithic sites (5th and 4th millennia bc) in Georgia (Lisitsina 1984) and from period VI beds in Tepe Yahya, Iran (Costantini and Costantini-Biasini 1985). If the identification and dating of these finds are correct, they rank among the oldest found yet. Furthermore, they come from sites situated not far away from the geographic belt where *P. miliaceum* might grow wild. Further east carbonized grains of broomcorn millet were discovered in several Yangshao culture farming villages (4th millennium bc) in the loess soil belt of north China (Ho 1977; An 1989). They occur together with remains of foxtail millet, an even older millet in China (see p.83).

Broomcorn millet appears rather early also in Europe. The earliest finds of the characteristically small oval seeds of *P. miliaceum* come from the late 5th and from the 4th millenia bc settlements in east and central Europe. Some of the oldest finds are from Danubian (Bandkeramik) sites in east Germany such as Eisenberg (Rothmaler and Natho 1957) or Leipzig (Körber-Grohne 1987, p. 333). Numerous carbonized grains are available also from 4130 bc Domica (= Kesevo) in Czechoslovakia (Fietz 1936; Tempír 1969). Somewhat later, lumps of pure charred grains were retrieved from the Funnel Beaker (TRB) beds in Szlachcin, Poland (Wasylikowa *et al.* 1991). Remains of *P. miliaceum* also occur in Tripolye culture sites (*c.* 3600 bc) such as Soroki in the Ukraine (Januschevich 1976), in Vinča culture contexts in Liubcova, Rumania (Comsa 1981); and in 3700–3600 bc Gomolava, Yugoslavia (van Zeist 1975). Broomcorn millet seems to have arrived in the Mediterranean basin from the north or the north-east (van Zeist 1980). In northern Italy, the millet appears in early Bronze Age (1700–1500 bc) settlements. Later it appears also in Ouroux-Marnay, central France (Hopf 1985). In Greece, it occurs in Kastanas and in Assiros Toumba, Macedonia (Fig. 23), but only in the late Bronze Age (Halstead and Jones 1980; Kroll 1983). A large quantity of charred grains of this crop was uncovered in Bronze Age beds (1900–1550 bc) at Haftavan, west Azerbaijan, Iran (Nesbitt and Summers 1988). Signs of the cultivation of *P. miliaceum* in the Near East appear late. The first known deposit of identifiable grains comes from 700 BC Nimrud, Iraq (Helbaek 1966b). There are so far no reports on *P. miliaceum* remains from the Indian subcontinent except for a few grains from 1600 bc Pirak, Pakistan (Costantini 1979).

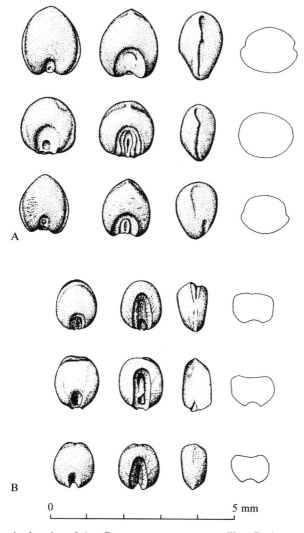

Fig. 23. Carbonized grains of A – Broomcorn or common millet, *Panicum miliaceum*. B – Foxtail millet, *Setaria italica*. Bronze Age Kastanas, Greece. (Kroll 1983.)

In conclusion, the evidence from the living plants and the archaeological remains is still fragmentary, but it does show that broomcorn millet is a relatively old crop, and that it does not belong to the Neolithic Near East crop assemblage. It seems as if *P. miliaceum* is a central Asian element which was picked up and added to wheat and barley agriculture soon after its arrival and establishment in these areas. But the scanty data available do not rule out the possibility that *P. miliaceum* (together with foxtail millet, see p. 83) represent an independent experiment in domestication, a process that could have started either in China or in central Asia prior to the arrival of the Near East grain crops.

Foxtail millet: *Setaria italica*

Foxtail or Italian millet, *Setaria italica* (L.) P. Beauv. (Fig. 24), is an appreciated minor crop in south-east Europe, several parts of Asia and North Africa. It is widely cultivated today in India, China, and

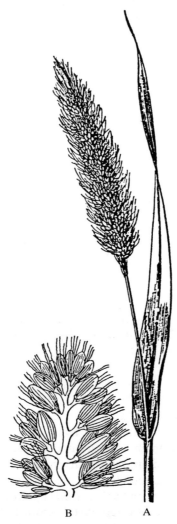

B   A

Fig. 24. Foxtail millet, *Setaria italica*. A – Fruiting panicle. B – Ripe branch of the panicle. (Hegi 1935, p. 268.)

Japan. This is a diploid ($2n$ = 18), predominantly self-pollinated millet with characteristic dense, bristle-bearing panicles and small, oval grains which are tightly enclosed by their pales. Similar to the broomcorn millet (p.78), foxtail millet is a warm season cereal. It survives well under dry conditions and completes its growth cycle in a short time.

The wild progenitor of foxtail millet is well identified. Cultivated *S. italica* shows close morphological affinities to and is interfertile with wild *S. viridis* (L.) P. Beauv., a common summer weed widely spread across Eurasia (De Wet *et al.* 1979) The spontaneous and the cultivated foxtail millets differ from one another mainly in their seed dispersal biology. Wild and weedy forms shatter their seed while the cultivars retain them. Genetic tests indicate that the shift is governed by two complementary recessive mutations (Rao *et al.* 1987).

Setaria italica is the principal crop of Neolithic agriculture in north China (Ho 1977; An 1989). Numerous remains of this cereal appear in the early farming villages that developed in the loess soil belt in the upper Yellow River basin. The earliest records come from the 6th millennium bc Peigang and Cishan contexts, i.e. the oldest farming cultures in north China (An 1989). Soon afterwards, *S. italica* emerges as the principal cereal of the Yangshao culture (5th and 4th millennia bc). The finds also demonstrate that already in Yangshao times foxtail millet was occasionally accompanied by common millet, *P. miliaceum* (p.80). Both millets continue to play a major role in China's food production also in more recent times (Ho 1977).

In Europe, carbonized seed that can be definitely assigned to *S. italica* first appears in the 2nd millennium bc Bronze Age settlements in central Europe (Netolitzky 1914) and in France (Hopf 1985). This millet is also reported from Late Bronze Age Kastanas, Macedonia, Greece (Kroll 1983; see Fig. 23). Additional finds are available from Iron Age Europe. The earliest definite evidence for foxtail millet cultivation in the Near East comes from Iron Age (*c.* 600 bc) Tille Höyük, south-east Turkey (Nesbitt and Summers 1988). Here a large quantity of pure grains was uncovered, indicating the use of foxtail millet as a crop.

In summation, the available archaeological evidence indicates that foxtail millet is a relatively old domesticant which very likely was first taken into cultivation in east Asia. Like common millet, *S. italica* does not belong to the Neolithic Near East crop assemblage. Moreover, it appears in Europe rather late. Since wild forms of foxtail millets are widely distributed across Eurasia, the evidence from the living plants contributes very little to the delimination of the place of origin of this crop. Better understanding of the start and spread of cultivated foxtail millet will necessarily depend on future archaeological finds.

Latecomers: sorghum and rice

Two additional principal cereals – sorghum, *Sorghum bicolor* (L.) Moench, and rice, *Oryza sativa* L. – are not native to the Near East and the Mediterranean basin. They are latecomers that arrived in these areas, as already fully developed crops, only in Greek and Roman times.

Sorghum

Sorghum bicolor is obviously an African domesticant (De Wet *et al.* 1976). The distribution of its wild races indicates that the crop must have been domesticated in the African savanna belt, south of the Sahara. Since the archaeological exploration of this part of Africa is still in its early stages, we have so far no critical data on the early stages of sorghum domestication. The earliest evidence of its cultivation comes from *c.* 2500 bc Hili in the Oman peninsula, and *c.* 1800 bc Pirak and several other 2nd-millennium bc sites in the Indian subcontinent (Cleuziou and Costantini 1980). These seem to represent the early migration of sorghum, from Africa via south Arabia, to India. The specialized *durra*-type which characterizes traditional sorghum cultivation in the Near East seems to have evolved in India and was transported from this subcontinent to the Near East (Harlan and Stemler 1976). The exact arrival of the *durra* material is not clarified yet; but definite signs of its cultivation in the Near East and the Mediterranean basin appear from Islamic times onwards.

Rice

Oryza sativa is a south- and east-Asiatic element. Wild and weedy races of this crop complex are widely distributed over the warm and wet parts of the Indian subcontinent, south-east Asia, Indonesia, and south China (Chang 1989). Because of this wide distribution and the extensive hybridization between wild types, weedy races, and cultivars, an accurate delimitation of the place of origin (on the basis of the living plants) is not feasible in Asiatic rice. The precise identification of the area – or areas – of domestication will have to wait till the archaeology of south and east Asia is better known than it is today.

In India, imprints of rice grains and rice husks were discovered in Neolithic Koldihava and Mahagra, near Allahabad and dated to the 5th and the 4th millennia bc. Yet it is not clear whether these finds represent cultivated rice or only wild rice (M. Kajale, personal communication). The earliest reliable finds of cultivated rice in India and Pakistan come from contexts dated to the second half of the 3rd millennium bc (Hutchinson 1976; Weber 1991, p.26). An early find is from Harappan culture Lathal, dated *c.* 2300 bc (Kajale 1991). In the 2nd millennium bc, rice seems to have been effectively added, as a summer crop, to wheat and barley cultivation in the north-western part of this subcontinent.

Even earlier rice remains are available from the area south of the Huai river and in the lower and central Yangtze regions of China (Ho 1977). They were radiocarbon-dated to the late 4th millennium bc and the 3rd millennium bc. Rice imprints are also reported from the lowest levels (*c.* 3000 bc) of Nom Nok Tha, Thailand (Solheim 1972).

Asiatic rice was probably introduced into the Near East in Hellenistic times. Both Greek and Roman writers knew about this cereal and said that it was grown in Baktria, Mesopotamia, and Syria (Lenz 1859, pp. 229–30). A large sample of rice grains was retrieved from a first century AD Parthian grave, Ville Royale II, Susa, Iran (Miller 1981). In Roman times a highly prized large kernel rice was produced in Israel (Feliks 1983).

3

Pulses

Pulses – annual legumes cultivated for their seed – accompany the cereals in most regions of grain agriculture. They are attractive first of all because contrary to most other flowering plants legumes are able to fix atmospheric nitrogen through symbiosis with the root bacterium *Rhizobium*. Rather than use nitrogen up, pulses add it to the soil. By rotation or mixing of legume crops with cereals the cultivator is able to maintain higher levels of soil fertility. Another virtue is that the seed of pulses are exceptionally rich in storage proteins whereas grass kernels are rich in starch. Thus they complement each other as food elements and contribute to a balanced human diet. In traditional agricultural communities pulses served – and still serve – as a main meat substitute.

Both the agronomic compensation and the dietary complementation between cereals and pulses have been appreciated in the early days of agriculture. Each major agricultural civilization developed not only its staple cereals, but also its characteristic companion legumes. Wheat and barley agriculture in west Asia and Europe had pea, lentil, broad bean, and chickpea. Maize in Meso-America was accompanied by *Phaseolus* beans; and in South America also by the ground nut. Pearl millet and sorghum cultivation in the African savanna belt were associated with cow pea and Bombara groundnut. Soybean was added to cereal cultivation in China, and hyacinth bean, black gram, and green gram in India.

Pulses seem to have started their role as companions of wheat and barley very early in the agricultural history of the Old World. The available archaeological evidence indicates that pea, lentil, chickpea and bitter vetch were taken into cultivation more or less together with the principal cereals. Their remains abound in the Near East Neolithic settlements. They are also common in the contexts of the Neolithic sites that appeared soon afterwards all over the vast area from the Atlantic coast of Europe to the Indian subcontinent. In other words, the accumulated archaeological data lead us to the conclusion that the establishment of food production in the Near East as well as the rapid spread of this technology from this 'core area' to west Asia and Europe were dependent not only on domestication of wheat and barley but also on the cultivation of legumes (Zohary and Hopf 1973; van Zeist 1980).

Several other pulses became crop species in west Asia and in Europe after the establishment of the 'first wave' of legume crops. Relatively early domesticants were the faba bean and the grass pea. They were followed, apparently much later, by fenugreek, and still later by the lupin. The cow pea, *Vigna unguiculata*, was domesticated in Africa south of the Sahara and reached the Mediterranean basin only in classical times.

In the seed legumes self-pollination also seems to have been a major asset in domestication. Similar to the cereals (p. 16), all Neolithic 'first wave' domesticates (pea, lentil, chickpea, and bitter vetch) are predominantly self-pollinated. So are their wild progenitors. Thus also in this group of crops, wild self-pollinated candidates were 'preadapted' for domestication as compared with cross-pollinated plants. The advantages conferred by self-pollination are the establishment of an barrier between wild and cultivated populations, and the automatic fixation of desired genotypes.

As in the cereals (p.18), evolution under domestication of the seed legumes is characterized by the development of a syndrome of domestication traits (Smartt and Hymovitz 1985, p. 45). Very probably many of these traits evolved as a result of unconscious selection (Zohary 1988).

First, there is an automatic selection for the retainment of seed in the pod. In most pulses (e.g. pea, lentil, the various members of the genus *Vicia*) seed dispersal in the wild depends on the bursting of the mature pods. The pods of cultivated forms do not burst or at least do not split open quickly. They are harvested and threshed by the cultivator. In both pea and lentil, the change to non-dehiscing or slow-splitting pods is a simple genetic event. It is governed by a single recessive gene.

A second conspicuous trend is the change in the dimensions and texture of the seeds. Most pulses under cultivation bear considerably larger seed than their wild progenitors. In some cases there is a three- or fourfold increase in the volume and the weight of the seed. This increase has been a gradual process. Seed from Neolithic contexts are still relatively small and not very different, in size, from those of their wild relatives. The large-seeded pulses with which we are familiar today reached their present dimensions only in classical times or even later. Another major change concerns the seed coat. Seeds of wild pulses have a thick and coarse testa which serves as a mechanical protection and for delay of germination. Smooth, thin seed coats evolved under domestication. The reduced seed coat is more permeable to water. A thin coat may thus reflect the breakdown of the wild-type germination regulation. Here, as in the cereals, the loss of germination inhibition seems to have been an automatic outcome of cultivation.

In several legumes domestication brought about striking changes also in the plant habit. The wild relatives of the cultivated pea, grass pea, and the vetches are climbers with tender branches and characteristic tendrils. Wild lentils are small, delicate plants. Under cultivation all these pulses evolved stiffer stems, a free-standing and robust habit, and a reduced dependence on climbing – traits that make them better adapted to grow in pure stands in tilled fields.

Finally, the wild-type chemical defences have been selected against. Many wild legumes contain potent toxins and antimetabolites in their seeds to protect them against animal predation. Cultivars of pulses frequently lack, or contain only reduced amounts, of these toxic compounds.

Lentil: *Lens culinaris*

Lentil ranks among the oldest and the most appreciated grain legumes of the Old World and today it is grown from the Atlantic coast of Spain and Morocco in the west to India in the east. In Mediterranean agriculture it is a characteristic companion of wheat and barley. Compared to the cereals, yields are relatively low (about 500–1500 kg/ha), but lentil stands out as one of the most nutritious and tasty pulses. The protein content is about 25 per cent and lentil constitutes an important meat substitute in peasant communities. Large quantitites of lentils are produced and consumed (soup, paste, in mixture with wheat or rice) in India, Pakistan, Ethiopia, the Near East, and the countries bordering the Mediterranean and further to the North.

The cultivated *Lens culinaris* Medik. (syn. *L. esculenta* Moench) manifests a wide range of morphological variation both in its vegetative and reproductive parts. Like many other annual grain crops, lentil is predominantly self-pollinated, and consequently numerous true breeding lines and aggregates of land races have evolved in this crop. All cultivated lentil varieties are diploid ($2n = 14$) and interfertile.

Conventionally, lentil cultivars are grouped in two intergrading clusters of seed sizes: (a) small-seeded lentils (subsp. *microsperma*), with small pods and small 3–6 mm seed; (b) large-seeded lentils (subsp. *macrosperma*), with larger pods and with seed attaining 6–9 mm in diameter. As in other pulses, domestication brought about the retention of the seed in the pod (non-dehiscence) and a gradual increase in seed size. *Macrosperma* forms are to be regarded as more advanced; they start to appear rather late in archaeological sequences – only in the 1st millennium bc. Carbonized seed is the main element in archaeological digs. Occasionally pods, or fragments of pods, occur as well.

Wild ancestry

Cultivated lentil belongs to the genus *Lens* Mill., a relatively small leguminous genus restricted in the wild to the Mediterranean basin and south-west Asia. In addition to the crop, the genus includes four annual, ephemeral, diploid ($2n = 14$) self-pollinated wild species. *Lens culinaris* shows close affinities to wild *L. orientalis* (Boiss.) Hand.-Mazz., a small pulse widely distributed in the Near East (Zohary 1972). In fact, *L. orientalis* (Fig. 25), looks like a miniaturized *L. culinaris* but bears pods that burst open immediately after maturation. Cytogenetic tests have shown (Ladizinsky *et al.* 1984) that *L. orientalis* is chromosomally variable. It contains a widespread chromosome type as well as several other forms which differ from the standard race by one, or even two, chromosome rearrangements. Chromosomally, the cultivated lentil is fully homologous with the standard *orientalis* race. Hybrids between such *orientalis* lines and *culinaris* cultivars are fully fertile. They show normal meiosis and regular formation of seven bivalents.

The cultivated lentil can be crossed also with *L. odemensis* Ladiz., a new taxon recently separated from *L. nigricans* and known up to date only from Israel and Turkey (Ladizinsky 1986). Yet this wild lentil is already quite distant from the crop. It differs from it by four chromosomal rearrangements; and the hybrids between them are largely sterile. Two other wild lentils: *L. nigricans* (Bieb.) Godr. and *L. ervoides* (Brign.) Grande are even more distant. Crosses between these wild lentils and the *culinaris–orientalis* complex usually end in hybrid embryo breakdown (Ladizinsky *et al.* 1984), indicating that both *L. nigricans* and *L. ervoides* are genetically quite distinct from the crop.

To sum up, the accumulated botanical and cytogenetic evidence in *Lens* establishes *L. orientalis* as the wild progenitor of the cultivated lentil. It also rules out the candidacy of other wild types in this small genus and indicates from what *orientalis* chromosome type the crop was derived. On the basis of this information *L. orientalis* should be considered as the wild stock of the crop complex. Its appropriate taxonomic ranking is therefore *L. culinaris* subsp. *orientalis*.

Like other grain crops, wild lentils differ from the cultivated varieties in their seed dispersal biology. As already mentioned, in *orientalis* plants the pods burst immediately after maturation. In contrast in cultivars the pods do not dehisce immediately and the seeds are retained. This difference is governed by a single mutation, the non-dehiscent condition being recessive to the dehiscent one.

Geographically, *Lens orientalis* is a Near-Eastern element (Map 7). It is distributed mainly over Turkey, Syria, Lebanon, Israel, Jordan, north Iraq, west and north Iran. It also penetrates into Afghanistan and adjacent central Asia. *Lens orientalis* grows primarily on shallow, stony soils, and gravelly hillsides in open or steppe-like habitats. It also enters disturbed localities such as stony patches or stone heaps bordering orchards and cereal cultivation. In most parts of its distribution, *L. orientalis* is rather inconspicuous or even rare. It usually forms small scattered colonies. However, on stony slopes of Mt. Hermon, the Anti-Lebanon, the oak park-forest belt of southern Turkey, and the western escarpments of the Zagros range, *L. orientalis* is occasionally locally common at 1200–1600 m altitude.

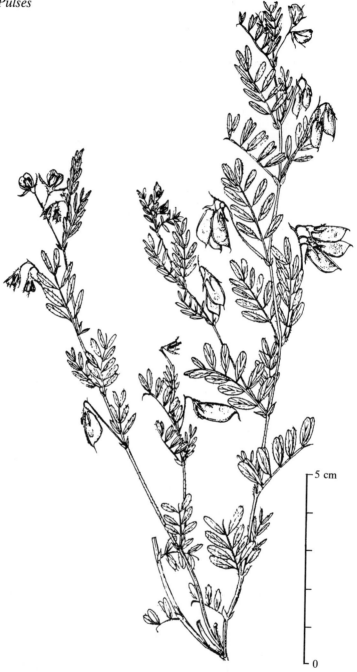

Fig. 25. Wild lentil, *Lens culinaris* subsp. *orientalis* (= *L. orientalis*). (M. Zohary 1972, plate 302.)

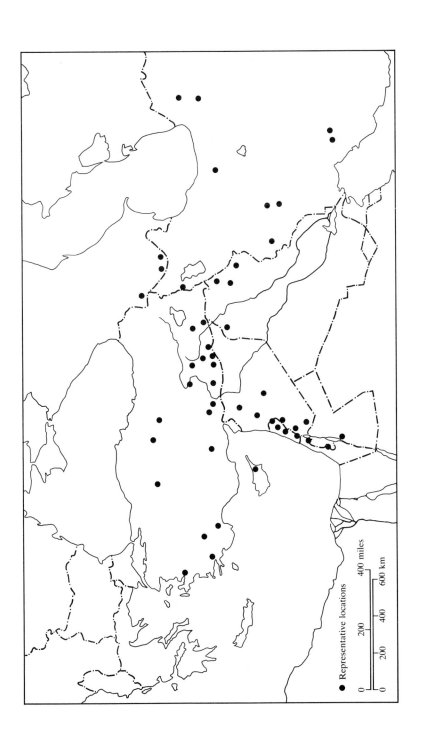

Map 7. Distribution of wild lentil, *Lens culinaris* subsp. *orientalis* (= *L. orientalis*). (After Zohary and Hopf 1973.) Wild lentil extends eastwards beyond the boundaries of this map into north Afghanistan and adjacent central Asia.

● Representative locations

0	200	200	400	400 miles
0	200	400	600 km	

Archaeological evidence

Lentils seem to be closely associated with the start of wheat and barley cultivation in the Near East. Very possibly this legume was domesticated in the region together with emmer, einkorn, and barley; that is, it should be regarded as a founder crop of Old World Neolithic agriculture (Zohary and Hopf 1973).

Lentils were apparently utilized in the Near East before the firm establishment of farming villages. Small, carbonized seed of lentil have been retrieved from pre-farming (9200–7500 bc) Mureybit (van Zeist 1970) and Tell Abu Hureyra (Hillman 1975), north Syria. In these early settlements we find the collection of wild *L. orientalis* together with wild einkorn wheat and wild barley. A few small lentil seeds have also been found in the Palaeo-and Mesolithic (pre-farming) layers of the Franchthi Cave, Greece (Hansen 1978, 1992) and in Grotta dell'Uzzo in Sicily (Constantini 1989); and it is very likely that they represent collected local wild *L. nigricans*.

Somewhat later, lentils appear in the aceramic farming villages that develop in the Near East arc during the 7th millennium bc (Map 2). A few small lentil seeds (2.5–3.0 mm in diameter) were detected in Jarmo, north Iraq (Helbaek 1959b), and the strata of Ali Kosh, Iran (Helbaek 1969). A single seed has been retrieved from aceramic Hacilar (Helbaek 1970); and a few more were dug out at Can Hasan, Anatolia (Renfrew 1968). Similar small lentils are reported from Tell Aswad (van Zeist and Bakker-Heeres 1985), Neolithic Tell Abu Hureyra. Syria (Hillman 1975) and Pre-Pottery Neolithic B Jericho (Hopf 1983) and in 'Ain Ghazal, Jordan (Rhollefson and Simmons 1985). Further to the north, lentils come from several levels in Çayönü, Turkey (van Zeist 1972). Evidently, lentils are present in most of the early (7th millennium bc) farming villages in the Near East, in which plant remains were carefully collected and analysed. They are small (2.5–3.0 mm in diameter) and usually do not occur in large quantities. Yet they are almost always present in company with cultivated wheat and barley.

A large hoard of carbonized lentils was recovered from the Pre-Pottery Neolithic B Yiftah'el, north Israel (6800 bc). The size of the hoard (*c.* 1 400 000 seeds) and its contamination by the fruits of the weed *Galium tricornutum* indicate that there and then lentil was already cultivated (Garfinkel *et al.* 1988). Large amounts of lentil seeds were discovered also in somewhat later phases of the Neolithic settlements in the Near East: in Tell Ramad, Syria, 6250–5650 bc (van Zeist and Bakker-Heeres 1985); in ceramic Hacilar, 5800–5000 bc (Helbaek 1970); in Girikihaciyan, 5000–4500 bc (van Zeist 1979–80); and in Tepe Sabz, Deh Luran Valley, Iran, 5500–5000 bc (Helbaek 1969). The Tepe Sabz lentils had already attained 4.2 mm in diameter. This is an obvious development under domestication.

In the 6th millennium bc lentils seem to be closely associated with the spread of Neolithic agriculture into south-eastern Europe. Here, too, they are found together with cultivated emmer, einkorn, and barley. Lentil-seed remains are present in almost all early Neolithic (6200–5300 bc) Greek agriculture settlements such as aceramic Ghediki (Renfrew 1979), the pre-ceramic basal levels of Argissa-Magula (Hopf 1962), Sesklo (Kroll 1981a), and Knossos (Renfrew 1979). Lentil remains frequent in later Neolithic and Bronze Age contexts in Greece such as Nea Nikomedeia (van Zeist and Bottema 1971) and Kastanas (Kroll 1983, 1984; Fig. 26). They were also found in some of the Early Neolithic (4800–4600 bc) Bulgarian sites such as Karanovo Mogila (Hopf 1973b; Renfrew 1979), as well as in Anza (Starčevo culture, 5300–4500 bc) in Yugoslavia (Renfrew 1976). In Hungary, lentils are reported from Neolithic Lengyel sites. In central Europe, they are present in numerous Danubian (Bandkeramik) settlements (Willerding 1980, table 4), frequently together with peas. As in the case of the pea (p. 101) and other pulses, lentils in Bronze Age settlements in Europe seem to be sparser than in Neolithic times. An increase in lentil finds is recorded in Iron Age settlements.

Lentils accompany wheats and barley also in the spread of the Near East agriculture southwards to Egypt and eastwards to the Caspian basin and the Indian subcontinent. The earliest records of *Lens* from Egypt come from Neolithic Merimde (M. Hopf, unpublished data). Lentils have also been found in pre-dynastic tombs dated 3500–3200 bc (Darby *et al.*

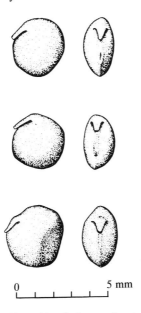

0   5 mm

Fig. 26. Carbonized seed of cultivated lentil, *Lens culinaris*. Bronze Age Kastanas, Greece. (Kroll 1983.)

1977) and at sites of all epochs ever since. Lentils are also recorded from Eneolithic Caucasia (Lisitsina 1984); and they are part of the Harappan crop assemblage in India and Pakistan and appear in Kashmir even earlier. (Kajale 1991)

In analysing lentil remains obtained from the early Near East sites, it is often difficult to decide whether they represent wild material or cultivated forms. The seed of wild *Lens* is morphologically very similar to that of the cultivated pulse. The only trait indicating domestication is the increase in seed size. Yet this process was slow and gradual. The first signs of seed size increase appear as late as in the end of the 6th millennium bc, i.e. some 1500 years after the definite establishment of wheat and barley cultivation in the Near East. We have, however, some circumstantial evidence suggesting that lentil cultivation is as old as agriculture itself. As already pointed out, the hoard discovered in 6800 bc Yiftah'el indicates cultivation. Indicative are also the data on the spread of Neolithic agriculture into Europe (Zohary and Hopf 1973). Lentils are repeatedly encountered in the early European agricultural settlements of the 5th millennium bc, situated far outside the areas in which either *L. orientalis* or *L. nigricans* grow wild. This is a strong indication that then and there lentils were already cultivated. Such early diffusion, together with the main Near Eastern cereals, could only be the outcome of an even earlier domestication, in the area where the wild progenitor occurs.

In summary, archaeological remains do not provide us with any direct diagnostic trait for a conclusive determination of the start of lentil domestication. Moreover, it is very doubtful whether comparative morphology will provide us with such clues in the future. Yet once Neolithic agriculture is soundly established, cultivation of lentil is part of it. The available archaeological information on early remains of lentil comes from the Near East. This is the very territory over which *L. orientalis* is distributed.

Pea: *Pisum sativum*

The pea ranks among the oldest grain legumes of the Old World. From its early beginnings, this crop was a close companion of wheat and barley cultivation (Zohary and Hopf 1973). Pea, *Pisum sativum* L., is well-adapted to both warm Mediterranean-type and cool temperate conditions. In peasant communities in the Near East, the Mediterranean basin, temperate Europe, Ethiopia, and north-western India, it constitutes an important source of protein for human consumption. The protein content of the seed is about 22 per cent. Today, pea ranks as the world's second most important pulse (Davies *et al.* 1985; Smartt 1990).

It is grown quite extensively in cool regions such as northern Europe and north-western USA. Mature dry seed was the principal product in classical times. Today, a substantial proportion of the crop is harvested as immature 'green' seed either for direct consumption or for canning and freezing.

Pea is a diploid ($2n$ = 14 chromosomes) and predominantly self-pollinated crop. As a consequence of the self-pollination system, variation in pea is moulded in numerous true breeding lines. These were used by Gregor Mendel for his classic genetic experiments, some one hundred and thirty years ago. Cultivated pea shows a wide range of morphological variation. Hundreds of land races are known and dozens of modern cultivars have been produced by breeders. All of them are interfertile. Varieties differ from one another in numerous traits such as the height and habit of the plant (from short field types to tall climbers), the colour of the flower (blue, purple, to white), the size and form of the seed, the texture and the colour of the seed coat, and the colour of the cotyledons (yellow or green). Forms with coloured flowers, relatively small seeds and long vines are frequently called field peas (var. *arvense*); while cultivars with white flowers, large seeds and shorter branches are known as garden peas (var. *sativum*). As in many other grain legumes, the conspicuous features of evolution under domestication are: non-dehiscence of the pod, i.e. the retention of seed in the ripe pod, a trait governed by a single recessive mutation; the gradual increase of seed size from 3–4 mm to 6–8 mm in diameter; and the reduction of the relatively thick texture of the seed coat, resulting in the breakdown of the germination inhibition of wild peas. Seed remains constitute the bulk of pea material recovered from archaeological digs. Other plant parts such as pods and vegetative material were hardly ever retrieved.

Wild ancestry

The cultivated pea belongs to a small leguminous genus, *Pisum* L., restricted in the wild to the Mediterranean basin and the Near East. All members of this genus are annual, diploid ($2n$ = 14) predominantly self-pollinated plants. On the combined basis of morphology, ecology, and cytogenetics, botanists recently revised the classification of peas, and recognized only two species in *Pisum* (Davis 1970).

P. sativum L. or the crop complex contains the variable aggregate of cultivated peas and the wild races closely related to the crop. All forms within the complex readily cross with one another and the hybrids are fully or almost fully fertile. All show full chromosome homology except for a single reciprocal translocation found in some wild forms and also in very few cultivated varieties.

P. fulvum Sibth. & Sm. is a distinct east-Mediterranean wild species with characteristic yellow-brownish flowers and chromosomes that are considerably divergent from those present in *P. sativum*. The two species are reproductively well isolated in nature, and their hybrids are semi-sterile.

The wild forms of *P. sativum* fall into two main morphological types:

(i)   A tall 'maquis type', previously called *P. elatius* M. Bieb. (= *P. sativum* L. subsp. *elatius* (M. Bieb.) Aschers. & Graebn.) is omni-mediterranean in its distribution and it thrives as a sporadic climber in maquis formations in the relatively mesic parts of the Mediterranean region. Occasionally *elatius* peas also colonize hedges and terraces bordering cultivation.

(ii) A shorter, more xeric 'steppe type' (Fig. 27), formally named *P. humile* Boiss. & Noë (= *P. syriacum* (Berger) Lehm; *P. sativum* L. var. *pumilio* Meikle), is geographically restricted to the Near East (Map 8). It occurs in the deciduous oak park-forest belt and in open, steppe-like, herbaceous vegetation characteristic to the Near East arc, i.e. in the same zone that harbours the wild progenitors of wheat, barley, lentil, and flax. From such primary habitats *humile* peas spill over to secondary habitats and occasionally infest cereal cultivation. In their general habit, some *humile* forms resemble the cultivated legume closely and differ from it mainly by their rough seed-coat and by pods that burst open and disperse the seed soon after maturation.

The boundaries between these two principal wild races are occasionally blurred and, particularly in Turkey, they are bridged by intergrading forms. Spontaneous hybridization between *humile* or *elatius* peas and the cultivated varieties also occur sporadically.

Chromosomally the wild peas of the *P. sativum* complex fall into two chromosomal types (Ben Ze'ev and Zohary 1973); *humile* material from the Mt. Hermon area and Turkey – and very probably also in other areas in the central and eastern parts of the Near East arc – have a chromosome complement identical to that prevailing in the cultivated varieties. Hybrids between such tame and wild forms show normal chromosome pairing and are fully fertile. In contrast, *humile* collections from south and central Israel and *elatius* material from Israel and Italy, differ from the crop by a single, reciprocal translocation. Hybrids in such cases show full chromosome pairing but somewhat reduced fertility due to the translocation heterozygosity.

To sum up, the evidence from living plants implicates the wild *humile* peas as the progenitor stock for pea domestication. *Humile* peas show closer morphological similarities than *elatius* peas to the cultivated aggregate

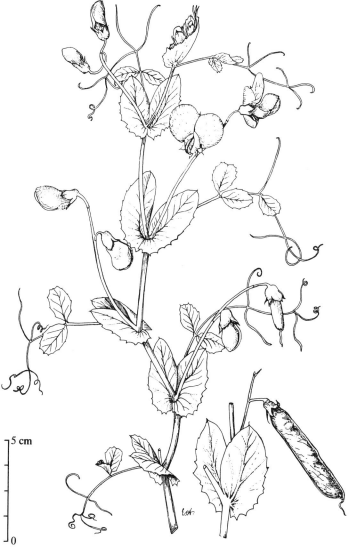

Fig. 27. Wild pea ('steppe-type'), *Pisum sativum* subsp. *humile* (= *P. syriacum*). (M. Zohary 1972, plate 317.)

and grow in steppes or steppe-like habitats, i.e. under open conditions that are not very different from those prevailing in the cultivated field. Within *humile* peas, the Turkish and Syrian forms having chromosomes identical with those present in the cultivars should be regarded as the primary ancestral stock. Yet it is very likely that the *humile* forms with the chromosome translocation, as well as the more mesic wild *elatius* peas,

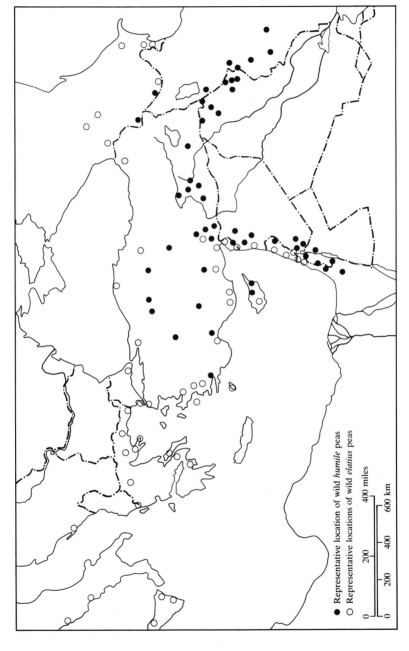

Map 8. Distribution of the two main wild races of pea, *Pisum sativum* (i) 'Steppe-type' *humile* forms and (ii) 'maquis-type' *elatius* forms. (After Zohary and Hopf 1973.) In the west Mediterranean basin wild *elatius* peas extend beyond the boundaries of this map and reach as far as Spain.

● Representative location of wild *humile* peas
○ Representative locations of wild *elatius* peas

0   200   400 miles
0   200   400   600 km

contributed some genes to the cultivated ensemble through occasional secondary hybridization.

Archaeological evidence

Peas are present in the Early Neolithic farming villages of the Near East (7500–6000 bc). Carbonized pea seed were discovered (see Map 2) in aceramic Jarmo, north Iraq (Helbaek 1959b), Çayönü, south-east Turkey (van Zeist 1972), Tell Aswad and Tell Ramad, south Syria (van Zeist and Bakker-Heeres 1985), and in the Pre-Pottery Neolithic B level in Jericho (Hopf 1983) and in 'Ain Ghazal, Jordan (Rollefson and Simmons 1985). Much richer remains of peas were available from somewhat later Neolithic phases in the Near East – from the 6th millennium bc. Large quantities of carbonized seed accompany the cultivated wheats and barley in 5850–5600 bc Çatal Hüyük (Helbaek 1964a), 5400–5000 bc Hacilar (Helbaek 1970) and 5800– 5400 bc Erbaba (van Zeist and Buitenhuis 1983).

In contrast to the wheats and barley, the earliest archaeological remains of peas do not provide us with simple traits for a foolproof recognition of cultivation (Zohary and Hopf 1973). In peas under cultivation there is a general trend towards an increase in the size of the seed and the length of the hilum, but such changes occurred gradually in the course of domestication, and in early finds, there is a considerable overlapping in the dimensions of wild and cultivated forms. Perhaps the most reliable indication of domestication in peas is provided by the surface of the seed-coat – as long as it is preserved. Wild peas are characterized by a rough or granular surface. Cultivated varieties have smooth seed-coats. However, in most of the carbonized seed samples, the seed-coat is missing, and it is impossible to ascertain whether the material retrieved represents wild or cultivated forms. The lower levels (7500–7000 bc) of Çayönü (van Zeist 1972) retained some fragments of seed-coats showing a rough surface, thus indicating a collection from the wild. Wild-type seed-coats occur even much later in Late Neolithic (5400–5050 bc) Hacilar (Helbaek 1970). Significantly, the remains from upper levels (*c.* 6500 bc) of Çayönü (van Zeist 1972), 5850–5600 bc Çatal Hüyük (Helbaek 1964a), and 5250 bc Can Hasan I (French *et al.* 1972) show the smooth seed-coat characteristic of domesticated varieties. This smoothness strongly suggests that cultivation of peas in the Near East is as old, or almost as old, as the cultivation of wheat and barley.

Peas seem to be associated with the spread of Neolithic agriculture into Europe (Zohary and Hopf 1973; van Zeist 1980). Here again they are closely associated with the wheat and barley production. Representative early sites in Greece include Early Neolithic Nea Nikomedeia, about 5500 bc (van Zeist and Bottema 1971), and the aceramic layers of Ghediki,

Sesklo, and Soufli (Renfrew 1966; Kroll 1981a). The Nea Nikomedeia carbonized seeds are well preserved and reveal the smooth coat of cultivated forms. Peas are further present in Early Neolithic Bulgaria. Finds from Tell Azmak (= Ašmaska Mogila) were dated to about 4330 bc (Hopf 1973b). Carbonized pea seed were also retrieved from Late Neolithic contexts in Gomolava (van Zeist 1975) and Valač (Hopf 1974), Yugoslavia.

Somewhat later pea is linked with the spread of the Danubian (Band-keramik) culture in central Europe (Willerding 1980; Knörzer 1991; Küster 1991), as well as the expansion of agriculture to Caucasia and Egypt. In the lower Rhine Valley, peas are common in sites dated 4400–4200 bc, and in central Germany large amounts of carbonized seed have been recovered (Baumann and Schultze-Motel 1968). In these samples, seed-coats are frequently well preserved and with one exception they show the smooth surfaces characteristic of the domestic crop. Also in Czechoslovakia peas appear already in Danubian (Bandkeramik) context (Wasylikowa *et al.* 1991).

The earliest peas from eastern Europe and Switzerland come from either late Neolithic or Bronze Age sites. The finds from Poland come from Funnel Beaker contexts in Ćmielów (Wasylikowa *et al.* 1991). A pure hoard of pea seeds was also found in Dudeşti culture bed in Cîrcea, Rumania (Cârciumaru 1991). So far there are no records of Neolithic peas in the Cardial culture sites in the west Mediterranean

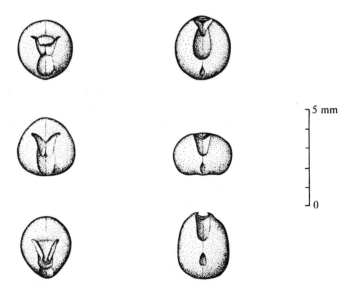

5 mm

0

Fig. 28. Carbonized seed of cultivated pea, *Pisum sativum* from Late Neolithic Dimini, Greece. (Kroll 1979.)

basin. The oldest specimens in this region come from Chalcolithic and Bronze Age layers. Finally, throughout Europe the Bronze Age finds of peas (as well as of other pulses) are fewer and sparser compared to the Neolithic finds. Richer deposits of pulses, including lentil and broad bean, occur again in the European settlements of later periods (Zohary and Hopf 1973).

Pea appears also in the Neolithic settlements of the Nile Valley and in Caucasia and Transcaucasia. In Egypt, early finds come from *c.* 5000 bc Merimde in the delta area (M. Hopf, unpublished data) and from prehistoric (3800–3400 bc) Nagada Khattara settlements in Upper Egypt (W. Wetterstrom, personal communication). *Pisum* is also present in contexts of the Shulaveri-Shomutepe culture (5th millennium bc to the beginning of the 4th millennium bc) in Arukhlo 1 and Arukhlo 2, Georgia (Lisitsina 1984; Schultze-Motel 1988). In India, the earliest finds to date come from *c.* 2000 bc Chirand, Bihar, and from several Harappa culture sites (Kajale 1991). In Kashmir, peas appear even earlier.

In conclusion, the archaeological evidence establishes pea as one of the earliest crops in the Near East Neolithic agriculture. Since this early start, pea seems to be a consistent element in Neolithic and Bronze Age food production throughout west Asia and Europe and a common companion of wheats and barley. The evidence from the living plants complements the archaeological finds. The wild *humile* forms, with chromosomes identical to those prevailing in the cultivated crop, should be regarded as the primary wild stock from which the cultivated were first domesticated.

Chickpea: *Cicer arietinum*

Chickpea is a valued seed legume of traditional agriculture in the Mediterranean basin, western Asia, including India and Ethiopia. It is a member of the grain ensemble found in Near East Neolithic and Bronze Age remains. Chickpea is adapted to a subtropical or Mediterranean-type climate; it grows almost exclusively in the post-rainy season on moisture stored in the soil. The main producers today are India with some 80 per cent of the total world production, the countries around the Mediterranean, and Ethiopia. Like lentil and pea, chickpea, with a seed protein content of some 20 per cent, constitutes an important meat substitute in peasant communities. Today this crop ranks third in the world's production of seed legumes.

Cultivated chickpea, *Cicer arientinum* L., is a predominantly self-pollinated annual crop with pods containing one to two seeds. The cultivars show a wide range of variation in size, colour, and shape of the seed, and in the size and form of leaves and flowers. All cultivated varieties are diploid ($2n = 16$ chromosomes) and interfertile. Chickpea land races are grouped into two interconnected clusters (Smithson *et al.* 1985; Smartt 1990, p. 241). Large-seeded varieties (known as 'Kabuli' type) with relatively smooth, rounded, light-coloured seed-coats and pale cream flowers predominate in the Mediterranean countries and the Near East. Varieties producing small, wrinkled seed ('Desi' type) with dark-coloured seed-coats and usually purple flowers prevail in the eastern and southern parts of the distribution area of the crop, i.e. in India, Afghanistan, and Ethiopia.

As in most other seed legumes, the conspicuous features of evolution under domestication in chickpea are the retention of pods and seed on the plants and the gradual increase in seed size from 3.5 to 6.0 mm and more. Large-seed forms are obviously more advanced. Another change is the development of a smooth seed-coat and the reduction of its thickness. Seed remains are the only material recovered up to date in archaeological digs.

Wild ancestry

The cultivated chickpea, *Cicer arietinum* L., is a member of a leguminous genus comprising some 40 species and centred in central and western Asia (Maesen 1972). The majority of *Cicer* species are herbaceous perennials or dwarf shrubs; some are annuals. The cultivated pulse shows close morphological affinities and an almost identical seed protein profile to two newly discovered wild species of chickpea: *C. echinospermum* Davis and *C. reticulatum* Ladiz. (Ladizinsky and Adler 1976a). These two wild chickpeas are diploid ($2n = 16$), self-pollinated annuals, known only from south-eastern Turkey. Externally they resemble each other closely, but they can be distinguished from one another by seed-coat texture which is echinate in *C. echinospermum* and reticulate in *C. reticulatum* and by their soil preferences. The first grows on basaltic soils while the second thrives on limestone bedrock. Of the two wild chickpeas only *C. reticulatum* (Fig. 29) is fully interfertile and contains chromosomes completely homologous to those found in the crop plant (Ladizinsky and Adler 1976b). Crosses between the cultivated chickpea and *C. echinospermum* and also those between *C. reticulatum* and *C. echinospermum* are difficult to achieve, and the interpsecific F_1 hybrids obtained proved to be highly sterile. Attempts to cross the cultivated chickpea with other wild *Cicer* species placed taxonomically close to *C. arietinum*, i.e. *C. pinnatifidum* Jaub. & Spach, *C. judaicum* Boiss. and *C. bijugum* Rech., failed altogether, indicating that these wild species are not related to the crop. The morphological and cytogenetic evidence available therefore implicates *C. reticulatum* in the ancestry of the cultivated plant. This wild chickpea should be regarded as the wild race of the cultivated crop and therefore referred to as *C. arietinum* subsp. *reticulatum*. Since the distribution of the wild progenitor is restricted

Fig 29. Seeds and ripe pods of wild chickpea *Cicer arietinum* subsp. *reticulatum*.

to south-eastern Turkey (Map 9), the area of origin of the cultivated crop can be outlined rather precisely. The central part of the Near East arc seems to be the territory in which chickpea was taken into cultivation.

Archaeological evidence

Like lentil and pea, chickpea seems to be closely associated with the start of food production in the Near East. Yet in Neolithic contexts in this region chickpea is more scantily represented than the other two pulses. Four carbonized chickpea seeds were recovered from aceramic (7500–6800 bc) Cayönü, Turkey (van Zeist 1972) and 15 more were found in the 8th millennium bc level of Tell Abu Hureyra, northern Syria (Hillman 1975). The seed still corresponds in size to that of *C. reticulatum* and, because these sites are situated within or close to the present distribution area of the wild progenitor, the finds could represent either wild or domesticated material. The seeds retrieved from Pre-Pottery Neolithic B (*c.* 6500 bc) Jericho (Hopf 1983) and 'Ain Ghazal (Rollefson and Simmons 1985); and those from the end of the 7th millennium bc level in Ramad near Damascus (van Zeist and Bakker Heeres 1985) very probably represent cultivated forms. The latter sites lie already far away from the territory of the wild progenitor, and the specimens from Jericho seem to have smooth coats. Early Bronze Age remains of chickpeas are more abundant. Considerable amounts of charred, well-preserved, fairly large seeds with smooth coats were retrieved in Israel and Jordan from Early Bronze Age Arad (Hopf

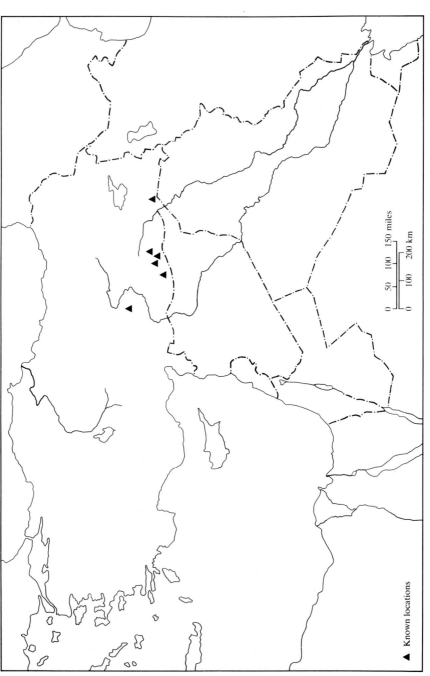

▲ Known locations

Map 9. Distribution of wild chickpea, *Cicer arietinum* subsp. *reticulatum* (= *C. reticulatum*). (After Ladizinsky and Adler 1976b.)

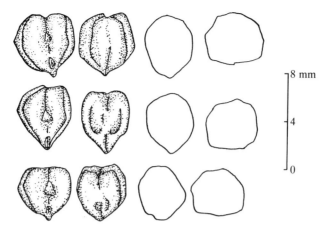

Fig. 30. Carbonized seed of cultivated chickpea, *Cicer arietinum*. Bronze Age Jericho. (Hopf 1983.)

1978a), contemporary Jericho (Hopf 1983; Fig. 30), and Bab edh-Dhra (McCreery 1979). These finds provide a clear indication of chickpea domestication since the only chickpea growing wild in Israel and Jordan, namely *Cicer judaicum*, has significantly smaller seed and a characteristic rough seed-coat.

In Greece, a single, well-preserved seed of chickpea was retrieved from Otzaki, Thessaly, from a pre-Sesklo layer, 6th millennium bc (Kroll 1981a). Richer finds are available from late Neolithic Dimini, *c.* 3500 bc (Kroll 1979). They measure 4.24 × 3.77 × 3.51 mm. The cultivated pulse must therefore have reached Europe with one of the first waves of migration of the Near East early grain crops.

The plant is missing in Neolithic layers of Yugoslavia and Bulgaria and the central and northern European countries. However, it appears in the west Mediterranean basin and has been reported from two early 3rd-millennium bc Chasséen sites in southern France (Courtin and Erroux 1974). If these finds are not intrusive, they represent an early arrival of this pulse to south France. Chickpea is also a member of the Neolithic Near East crop assemblage present in Harappan settlements in the Indian subcontinent (Vishnu-Mittre 1977; Kajale 1991). From Bronze Age to classical times, chickpea evidently ranked among the much consumed pulses of the Mediterranean basin and the Near Eastern countries.

In conclusion, the evidence from the living plants and the plant remains discovered in archaeological digs indicate that *Cicer arietinum* belongs to the Early Neolithic grain crop assemblage of the Near East. Archaeological data are still limited; but in this legume, the delimitation of the place of origin is relatively simple: the wild progenitor of the cultivated chickpea is endemic to the central part of the Near East arc. Here, very likely, this pulse was first brought into cultivation.

Faba bean: *Vicia faba*

Together with lentil, pea, and chickpea, the faba bean, *Vicia faba* L., (also known as broad bean or horse bean), belongs to the principal pulses of Old World agriculture. It grows well in both warm, summer-dry Mediterranean environments and in the more northerly temperate parts of Europe and Asia (Bond *et al.* 1985). The erect, robust plant bears readily threshed pods and relatively large seeds of high protein content, about 20–25 per cent. In some Asian and Mediterranean countries, particularly in Egypt, the dry seed of the faba bean provides the principal protein source for the poor. But the pods are also used when green; moreover, in both Europe and Asia the seed is regarded as a valuable animal feed, hence also the name 'Horse bean'. Large quantities of *V. faba* are produced in several Mediterranean countries: Egypt, Morocco, Spain, Italy, Turkey, as well as in Ethiopia, temperate Europe, western Asia, and especially in China, which is at present the main producer.

Vicia faba is a diploid plant with $2n = 12$ chromosomes. In contrast to the majority of the grain crops, faba bean is not a self-pollinated crop (Lawes 1980). Although the crop is self-compatible, a considerable amount of cross-pollination takes place in most cultivars. Furthermore, when subjected to selfing many varieties manifest various degrees of inbreeding depression. Some faba beans, particularly *paucijuga* forms from India and Afghanistan, are, however, able to self-pollinate.

Under domestication, faba beans have developed a considerable amount of morphological variation and different ecological adaptations. Conspicuous intraspecific differences are displayed in various traits like vegetative habit, pod structure, pod shattering, and the shape, colour, and size of the seed. Seed size has served as the principal character for intraspecific subdivision. Faba bean taxonomists (Muratova 1931; Hanelt 1972) recognize three or four intergrading and interfertile main types in *V. faba*. Forms with relatively small rounded seed measuring 6–13 mm are placed in *V. faba* var. *minor*. Some small-seeded forms, grown mainly in India, Pakistan, and Afghanistan show reduced numbers of leaflets (mostly one pair per leaf) and have very small flowers. They are frequently regarded as a distinct intraspecific taxon: var. *paucijuga*. Forms with medium-sized seed, that are 15–20 mm long, 12–15 mm wide, and 5–8 mm thick, are placed into var. *major*. Large-seeded faba beans evolved very late under domestication. Archaeological remains ranging from the Neolithic period to Roman times are all within the range of var. *minor*.

Wild ancestry

The wild ancestor of the cultivated faba bean, *V. faba* L., has not yet been discovered (Zohary 1977). The cultivated pulse (with $2n = 12$ chromosomes) shows close morphological similarities to a group of closely related Mediterranean and Near East wild vetches: *V. narbonensis* L., including types known as *V. serratifolia* Jacq. and *V. johannis* Tamamsch.; and *V. galilea* Plitm. & Zoh., including *V. hyaeniscyamus* Mouterde. Taxonomically, all these wild vetches are placed together with the crop in the section *Faba* of the genus *Vicia* L. But in spite of the striking morphological resemblance, particularly between *V. faba* and *V. galilea*, the crop is chromosomally distinct. Extensive cytogenetic tests carried out in the section *Faba* (Schäfer 1973) revealed that all known wild types contain not $2n =12$ but $2n = 14$ chromosomes. Repeated attempts (Schäfer 1973; Ladizinsky 1975) to cross the crop with any of the known wild species of section *Faba* have failed altogether, indicating that *V. faba* is not only chromosomally unique but that it is also reproductively strongly isolated from all known wild members of this section. This rules out the candidacy of the *narbonensis-galilea* $2n = 14$ vetches from the ancestry of the crop. Consequently, in the case of the faba bean we still need to discover an elusive 12-chromosomed wild progenitor which, in all likelihood, is very limited in its geographic distribution. There is also the possibility that the wild ancestor became extinct.

Archaeological evidence

The earliest remains of faba beans come from Pre-Pottery Neolithic B (6800–6500 bc) Yiftah'el near Nazareth, Israel. Here a hoard of some 2600 well-preserved charred seeds was discovered (Kislev 1985). The seed are small ($5.5 \times 4.7 \times 4.0$ mm) but distinctively flat. They are also thickest at the hilum's side. These features indicate that they may belong to *V. faba* and not to wild *V. narbonensis* or *V. galilea*. The latter have more globular seed.

Apart from this find there is no other unequivocal record of faba beans in the Neolithic farming villages in the Near East. Very few seeds resembling *faba* were found in Pre-Pottery Neolithic B Jericho (Hopf 1983), aceramic Tell Abu Hureyra, Syria (Hillman 1975) and Cape Andreas Kastros, Cyprus (van Zeist 1981). But these scanty remains do not permit us to decide whether one is faced with the cultivation of faba beans or with the collection of non-related wild species belonging to the section *Faba* such as *V. narbonensis* or *V. galilea*. Similarly rare remains of *V. faba*-type seed appear in Late Neolithic Sesklo and Dimini in Greece (Renfrew 1966; Kroll 1979). Again the seeds are small and globular. They conform well also with wild *V. narbonensis*.

Larger quantities of frequently somewhat bigger seed appear in final Neolithic/Bronze Age settlements (3rd millennium bc) in the west Mediterranean basin, and in the contemporary Bronze Age sites in south and central Europe, and in the Near East (Map 10). The G. Schweinfurth collection in Berlin contains *V. faba* seed found in a 5th-dynasty tomb in Egypt. Similar finds continue to appear in the 2nd millennium bc. The available remains are clustered in the following main areas (for reviews see van Zeist 1980; Hopf 1991a):

(i) the Iberian peninsula – both south Spain and Portugal (Fig. 31);

(ii) north Italy and Switzerland – with an extension to Austria, Czechoslovakia, and Germany;

(iii) the Aegean belt with representative finds in Bronze Age Kastanas (Kroll 1983) and Lerna (Hopf 1961b);

(iv) the east Mediterranean region including Bronze Age Arad (Hopf 1978a) and Jericho (Hopf 1983).

Significantly, many of these finds, particularly from central Europe, lie already outside the range of the 14-chromosomed wild members of section *Faba*, ruling out the possibility that the excavated seeds belong to them. Furthermore, seed size in several locations is already larger than that found in the wild vetches. This strongly suggests that then and there we are already confronted with cultivated *V. faba*. Numerous finds of faba beans are available from Iron Age and classical times in Europe and west Asia, indicating that *V. faba* has been used as a major food source.

When the evidence from the living plants and from the archaeological remains is put together we are led to the following conclusions:

(i) The wild progenitor of the cultivated faba bean is not yet known and we still need to discover the elusive 12-chromosome ancestor.

Archaeological sites in Map 10
1: Jericho (Hopf 1983). **2:** Arad (Hopf 1978). **3:** Beit-Shan (Feinbrunn 1938). **4:** Yiftah'el (Kislev 1985). **5:** Tell Abu Hureyra (Hillman 1975). **6:** Sahure (see Darby *et al.* 1977, p.682). **7:** Lerna (Hopf 1961b). **8:** Sesklo (Renfrew 1966). **9:** Dimini (Kroll 1979). **10:** Kastanas (Kroll 1983). **11:** Hissarlik (Buschan 1895, p.213). **12:** Burgschleinitz (Werneck 1949, p.251). **13:** Passo di Corvo (Follieri 1973). **14:** Luni (Helbaek 1976). **15:** Monte Leone (Pals and Voorrips 1979). **16:** Several sites in northern Switzerland (see Hopf 1970). **17:** Several sites in west Switzerland (see Hopf 1970). **18:** Lac de Bourget (Buschan 1985, p.213). **19:** Grotte Murée (Courtin and Erroux 1974). **20:** Almizaraque (Netolitzky 1935; M. Hopf, unpublished data). **21:** El Argar (Buschan 1895, p. 213; Hopf 1991b). **22:** Cueva del Toro (Hopf 1991a). **23:** Zambujal (Hopf 1981a). **24:** Vila Nova de S. Pedro (do Paço 1954). **25:** Pepin/Amarante (do Paço 1954).

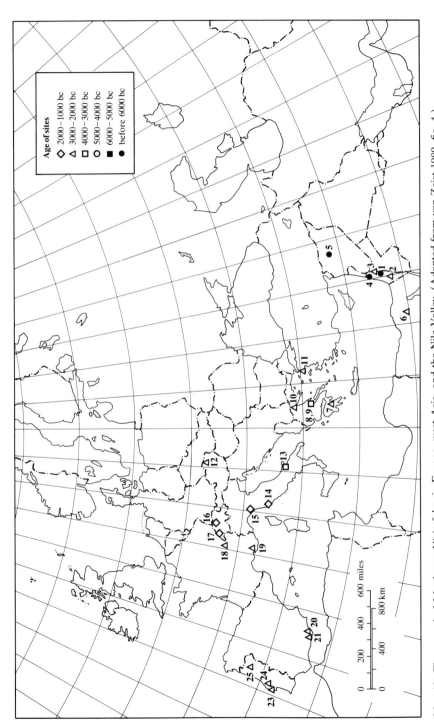

Map 10. The spread of faba bean, *Vicia faba*, in Europe, west Asia, and the Nile Valley. (Adapted from van Zeist 1980, fig. 4.)

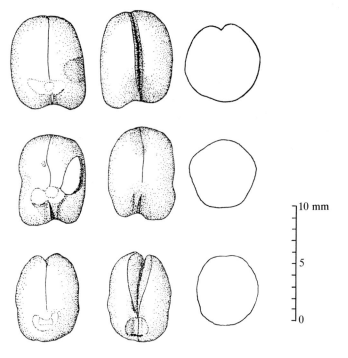

Fig. 31. Carbonized seed of faba bean, *Vicia faba* var. *minor*. Early Bronze Age Zambujal, Portugal. (Hopf 1981a.)

(ii) Numerous remains of faba beans appear, rather suddenly, in various parts of the Mediterranean basin and central Europe in the 3rd millennium bc.

(iii) We still know very little about the beginnings of *V. faba* domestication. The 7th millennium bc remains discovered in north Israel indicate association with the early Near East crop assemblage. More definite conclusions could only be reached after additional archaeological evidence is recovered and/or when the wild progenitor of the crop is found.

Bitter vetch: *Vicia ervilia*

Bitter vetch, *Vicia ervilia* (L.) Willd. is a small, diploid ($2n = 14$), self-pollinated pulse with beaded pods and characteristic angular seed. Today it is grown as a useful minor crop in the Mediterranean basin and the Near East; the principal producer is Turkey. As its name implies, its seeds are bitter and toxic to humans and to some animals. But the poisonous substance can be removed by soaking in water (van Zeist 1988). At least since Roman times this vetch has been utilized primarily as an animal feed, for it is regarded as very inferior for human consumption and is only eaten by the very poor, or in times of famine.

The wild ancestry of the cultivated bitter vetch is satisfactorily established (Zohary and Hopf 1973; Ladizinsky and van Oss 1984). Truly wild forms of this vetch, growing in primary habitats, are rather restricted in their distribution and known from Anatolia, north Iraq, the Anti-Lebanon (including Mt. Hermon) and from Jebel Druz (Map 11). They show a striking resemblance to the cultivated crop but differ from it by their dehiscent pods and slightly smaller seeds. Hybrids between the cultivars and the wild forms are fully fertile. Weedy races and feral forms of *V. ervilia* occasionally infest grain crops and edges of cultivated fields throughout the Near East and Greece.

Seeds of bitter vetch first appear in several agricultural settlements in Turkey of the 7th and 6th millennia bc. Large amounts of carbonized seed of *V. ervilia* were found in various phases of aceramic (7500–6500 bc) Neolithic Çayönü (van Zeist 1972), but it is impossible to determine whether these plant remains represent wild or domesticated material. Bitter vetch is also common in aceramic Can Hasan III (Renfrew 1968; French *et al.* 1972). Somewhat later, it appears in contexts of the 6th millennium bc Çatal Hüyük (Helbaek 1964a), Hacilar (Helbaek 1970), and Erbaba (van Zeist and Buitenhuis 1983). It also occurs in remains of 5th millennium bc Girikihaciyan (van Zeist 1979–80).

Considerable amounts of carbonized seed of *V. ervilia* have been discovered in late Neolithic and Bronze Age Greece and even larger quantities in remains of both these periods in Bulgaria. The earliest rich and well identified bitter vetch finds come from the beds of middle 6th millennium bc Nea Nikomedeia (van Zeist and Bottema 1971) and they are followed by several late Neolithic and early Bronze Age finds (Renfrew 1979; Kroll 1983). In Bulgaria huge, pure hoards of bitter vetch grains were discovered in Middle Neolithic Azmaška Mogila (Hopf 1973b). Carbonized remains of this pulse are also very frequent in Bulgaria throughout the Eneolithic and the Bronze Age (Renfew 1979; Janushevich 1978) and sometimes constitute the main plant material retrieved from the sites.

Outside Turkey, Greece, and Bulgaria, *V. ervilia* is much less common. Over large areas it is missing altogether. A few grains of bitter vetch were retrieved from Late Neolithic Gomolava, in Serbia (van Zeist 1975), Cascioărele, Rumania (M. Hopf, unpublished data) and also from the early Tripolye culture site of Karbuna in Moldavia (Janushevich 1978). But on the whole remains of *V. ervilia* in Neolithic and Bronze Age contexts show a distinct regional pattern: the cultivation of this pulse in the past seems to have been heavily centred in Turkey, Greece, and Bulgaria.

The combined evidence from the living plants and from the archaeological remains indicates that bitter vetch was part of the Near East Early

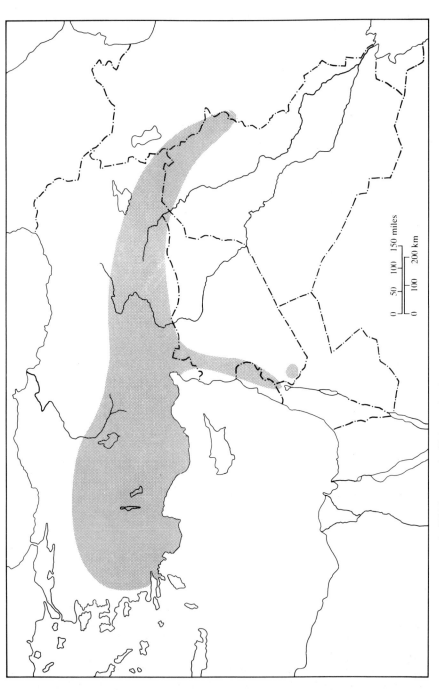

Map 11. Distribution area of wild bitter vetch, *Vicia ervilia.* (Based on Zohary and Hopf 1973; Townsend 1974; Ladizinsky and van Oss 1984.)

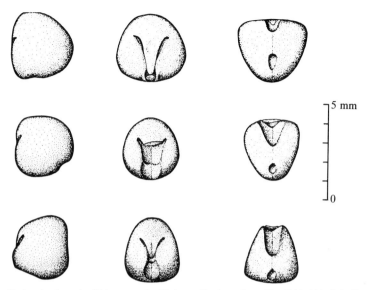

5 mm

0

Fig. 32. Carbonized seed of bitter vetch, *Vicia ervilia* from Late Neolithic Dimini, Greece. (Kroll 1979.)

Neolithic crop assemblage. This pulse was evidently taken into cultivation in Anatolia or the Levant, i.e. in the general area in which it still grows wild today. However, since there are no reliable diagnostic traits by which wild and cultivated forms of bitter vetch in archaeological remains can be distinguished from each other, the early Turkish archaeological finds could be either. The largeness and purity of the numerous samples of this pulse retrieved from the Neolithic and Bronze Age sites in the Balkans and Turkey strongly suggests, though, that *V. ervilia* was already cultivated at that time. We know very little about the mode of utilization of this bitter seeded legume by the Neolithic and Bronze Age farmers.

Common vetch: *Vicia sativa*

Common vetch, *Vicia sativa* L., is another member of the genus *Vicia* which characterizes Mediterranean grain agriculture. It is a minor crop cultivated for hay and for seed. Similar to bitter vetch, the seeds are not attractive for human consumption. The pulse is used today exclusively as an animal feed. Common vetch is also a frequent contaminant of lentil and bitter vetch cultivation. Seed of the latter pulses sold in local markets frequently contain scattered *V. sativa* seed.

The cultivated *V. sativa* is a diploid self-pollinated plant (chromosome number $2n = 12$) with straggling or ascending habit and rounded, somewhat compressed, smooth seeds, 4.5–7.0 mm in diameter. The cultivars are closely related to an extraordinarily variable (and chromosomally complex) aggregate of wild types and weedy forms, the distribution of which is centred in the Mediterranean basin (Zohary and Plitman 1979). All are now grouped, together with the crop, in the *V. sativa* complex. Most cultivars, together with morphologically closely related weeds and escapees are placed in *V. sativa* subsp. *sativa*.

Carbonized seeds of *V. sativa* have been reported from several Neolithic and Bronze Age sites in the Near East and Europe. But since seed sizes of weedy forms and wild types overlap considerably those found in the cultivars, it is difficult to conclude whether the remains represent cultivated forms, weedy contaminants or collection from the wild. More definite indications of common vetch cultivation are available only from Roman times.

The earliest archaeological records of *V. sativa* come from Natufian and Neolithic Tell Abu Hureyra, Syria (Hillman 1975) and from preceramic Neolithic Can Hasan III, Turkey (French *et al.* 1972). They are followed by several records from Neolithic and Eneolithic Bulgaria (Renfrew 1973, p. 188), Hungary (Hartyányi and Nováki 1975) and Slovakia (Hajnalová 1975). Common vetch is also reported from several Bronze Age contexts such as the second half of the 3rd millennium bc beds in Ak-Tepe near Ashkabad, Turkmenia (Priščepenko 1973, as cited by Schultze-Motel 1974) and from Slovakia (Kühn 1981).

Grass pea: *Lathyrus sativus*

Grass pea or chickling vetch, *Lathyrus sativus* L., is another minor pulse crop of traditional agriculture in the Mediterranean basin, south-west Asia, Ethiopia, and north-western parts of the Indian subcontinent. India is the main producer today, and in this country *L. sativus* is appreciated for its ability to grow in dry places and on poor soils. Grass pea is a diploid ($2n = 14$), self-pollinated annual with branched, straggling, or climbing habit, blue (sometimes violet or white) flowers and characteristic smooth seed with pressed sides. At present the pulse is used mainly as an animal feed, though in India it serves also as an article of human diet and is consumed by the very poor, and in times of famine.

Cultivated grass pea shows close morphological resemblance to a group of wild *Lathyrus* species distributed over the Mediterranean basin and south-west Asia. It is closest to *L. cicera* L., a wild grass pea growing in several East Mediterranean and Near Eastern countries (Greece, Turkey, North Iraq, North Iran, Transcaucasia). In these countries, *L. cicera* also abounds as a weed in cereal cultivation. Very likely *L. cicera* is the wild progenitor of the cultivated grass pea. But since several other *Lathyrus* species (*L. marmoratus* Boiss & Bl., *L. blepharicarpus* Boiss.

and *L. pseudocicera* Pamp.) also show close morphological affinities with *L. sativus*, the final determination of ancestry in this crop has to depend also on critical cytogenetic tests. These, however, are as yet unavailable. As in many other grain legumes, domestication of the grass pea resulted in an increase in seed size. Seed of *L. sativus* cultivars are somewhat larger (6–8 mm in diameter) than those of their wild relatives (5–6 mm).

Carbonized seeds of grass pea appear in several Near East, Aegean, and west Mediterranean Neolithic settlements. Few grass pea seeds were discovered in 7th millennium bc Jarmo, Iraq (Helbaek 1960a) and Çayönü, Turkey (van Zeist 1972). It is impossible to decide whether they represent cultivated forms or collection from the wild. Recently a flotation sample from final Pre-Pottery Neolithic B beds in Gritille, a site on the Euphrates in south-east Turkey, yielded over 800 carbonized seeds of grass pea (Miller 1991). The bulk of Neolithic grass pea finds comes, however, from 6th and 5th millennia bc Greece and Bulgaria. About 200 seeds were discovered in *c.* 6000 bc Prodromus (Halstead and Jones 1980), a 7 litre hoard was uncovered in 5th millennium bc Servia (Hubbard 1979), and some more in Late Neolithic Dimini (Kroll 1979). Numerous seeds were also retrieved from 4700 bc Azmaška Mogila, Bulgaria (Hopf 1973b). Few *Lathyrus*-type carbonized seeds were uncovered also in several Impressed Ware (Cardial) and Chasséen sites in south France (Courtin and Erroux 1974).

In Bronze Age, remains of grass pea continue to appear in the Near East and in south-eastern Europe (for a compilation of data see Kislev 1989). They were discovered, for example, in Early Bronze Age Lachish, Isreal (Helbaek 1958), in Tell Basmosian, Iraq (Helbaek 1963), in Late Bronze Age Kastanas, Greece (Kroll 1983; see also Fig. 33), and in Middle Bronze Age Tiszaalpár Várdomb, Hungary (Hartyányi 1982). In most of these finds the seeds are small and the seed-coats are missing. It is therefore impossible to decide whether they represent a collection from the wild, an infestation of *L. cicera*-type weeds in fields of other crops, or cultivated forms. Yet the sheer quantities of the finds in Gritille, Prodromus, Servia, Azmaška Mogila, and Dimini seem to indicate cultivation.

In conclusion, the grass pea also seems to belong to the Early Neolithic crop assemblage; or it was added to it soon after the establishment of grain agriculture. When and where this happened is still difficult to say, more so since the wild relatives of this pulse are widely distributed over the entire Mediterranean basin. The recent find in Neolithic Gritille suggests that *L. sativus* may have been taken into cultivation in the Near East; yet the proposal of south Balkan origin (Kislev 1989) is another valid option. More archaeobotanical information is needed to clarify the time and place of origin of this pulse.

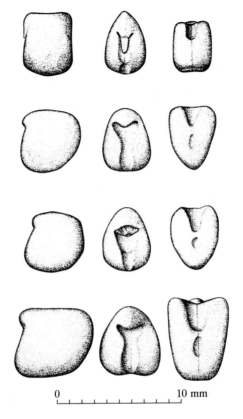

0   10 mm

Fig. 33. Carbonized seed of grass pea, *Lathyrus sativus*. Bronze Age Kastanas, Greece. (Kroll 1983.)

Fenugreek: *Trigonella foenum-graecum*

Fenugreek, *Trigonella foenum-graecum* L., is a useful minor pulse crop in traditional agricultural communities in the Mediterranean basin, south-west Asia, Ethiopia, and the northern parts of the Indian subcontinent. The seed is widely used as a condiment and as an important pulse ingredient for the preparation of curries and soups. Fenugreek is also used today as an effective soil renovator. It is a diploid ($2n = 16$), self-pollinated, annual legume with long, linear-lanceolate, stiff pods and characteristic roughly quadrangular seeds which have very prominent radicles or 'beaks'.

The crop shows very close morphological resemblance to a group of wild *Trigonella* species distributed over the east Mediterranean basin and the Near East (Širjaev 1932). It is not yet clear which member of this series (section *Foenum graecum* Širjaev) gave rise to the cultivated fenugreek. But since all these wild relatives have rather small distributional areas and are restricted to the Near East and the eastern parts of the Mediterranean basin it seems very likely that *T. foenum-graecum* was brought into cultivation somewhere within this region.

Carbonized fenugreek seed is only available from a few Near Eastern sites: from 4000 bc Tell Halaf, Iraq (Neuweiler 1935), from Early Bronze Age Lachish, Israel (Helbaek 1958), and from 3000 bc Ma'adi (Neuweiler 1946), and fourteenth century BC Tutankhamun tomb (Germer 1989a) in Egypt. All in all, the available archaeological evidence on this pulse is still fragmentary; yet it indicates that fenugreek was already part of the Near East agriculture in the Early Bronze Age.

Lupins: *Lupinus*

Several annual members of the genus *Lupinus* L., distributed over the Mediterranean basin have been taken into cultivation and are grown today either as grain crops or as forage plants and green manure. Prominent among them are the white lupin, *L. albus* L., the yellow lupin, *L. luteus* L., and the narrow-leaved lupin, *L. angustifolius* L. (for details see Duke 1981). All are vigorous growers and produce large, attractive seed. Yet their use is complicated by the fact that lupins generally contain bitter alkaloids which are difficult to remove. Domestication of these pulses meant first of all selection for non-bitter seed.

The white lupin, *L. albus* (syn. *L. termis* Forssk.) at least has been cultivated in the Mediterranean countries since antiquity and should be regarded as characteristic pulse element of this region. Boiling or steeping in water removes the bitter alkaloids from its white-coated seed and makes them edible.

The wild progenitor of the cultivated white lupin is well identified. The cultivars are interfertile and chromosomally identical ($2n = 50$) with wild forms native to the Aegean region. These were traditionally called *L. graecus* Boiss. & Spruner. More recently they were included in the crop complex as *L. albus* subsp. *graecus* (Boiss. & Spruner) Franco & Silva. Wild forms are characterized by dehiscent pods, smaller, more bitter seed and thicker, pigmented seed-coats.

Well preserved and easily identifiable seed of cultivated *L. albus*, with the characteristic smooth white coats, are available from Roman Egypt (Darby *et al.* 1977). Together with a single find reported from Bronze Age Hala Sultan Tekke, Cyprus (Hjelmqvist 1979b) they provide the earliest indication of white lupin domestication.

Charred, well preserved lupin seeds (yet without the seeds-coats) were found in the Natufian beds of Hayonim Cave, Israel (Hopf and Bar Yosef 1987). These remains represent collection from the wild of a local lupin species.

4

Oil and fibre crops

Plants producing oil and/or fibre were also taken into cultivation in the early stages of agriculture in the Old World. Flax is apparently the earliest and the best documented oil and fibre crop. Remains of linseed and fragments of linen indicate that flax belongs to the 'first wave' domesticates of the Near East and that it maintained its leading role in the Old World's fibre and oil production from Neolithic times to the early years of the twentieth century. Two other important fibre crops are hemp and cotton. The evidence for the beginnings of their domestication is still insufficient; but it is clear that they were introduced into cultivation outside the Near East core area. When this happened is hard to say; yet both hemp and cotton were apparently already used in pre-classical times.

Several oil plants seem to have entered cultivation not directly, but first by evolving weedy forms. They were added to the crop assembly only after the firm establishment of the principal seed crops such as wheat, barley, and flax. The case of the false flax, *Camelina sativa* (p.131) is well documented. Several other cruciferous plants (e.g. *Sinapis* and members of the genus *Brassica*) seem to have followed a similar mode of evolution under domestication.

In a similar way to the cereals and the pulses, domestication of oil-bearing seed plants triggered the evolution of characteristic traits. Also these crops are maintained by sowing, reaping, and threshing. Under such conditions, unconscious selection will have operated to bring about the breakdown of the wild mode of seed dispersal and the retention of the seed on the mother plant. Indeed, in most oil crops the fruits do not dehisce. In others there is a delay in fruit opening. Another conspicuous evolutionary trend is yield increase. Compared with their wild relatives, most cultivars of oil bearing seed crops have larger inflorescences and/or bigger fruits (with larger or more numerous seeds). The wild mode of germination inhibition has also broken down; and seed coats have become thinner. Finally, cultivated varieties frequently show increased oil content. Yet this is not a general rule. Seeds of some wild types are as rich in oil as their cultivated counterparts. In fibre-producing plants the accent is on long, strong fibres. If stem fibres are used (in flax and hemp), long and straight plants evolve. If lint is the goal (in cotton), mutations determining long lint (not found in the wild) are selected for.

We still know very little how oil and fibres were extracted from plants in the early phases of agriculture. However, it sounds reasonable to assume that

(i) Oil could be obtained by decantation, i.e. by crushing the seed, pouring hot water on the meal, and scooping the oil after setting. In some cases, before crushing, the seed were softened by allowing them to absorb water.

(ii) Flax and hemp stems were retted in order to free the fibres.

Very possibly both technologies antedate agriculture.

Flax: *Linum usitatissimum*

Flax, *Linum usitatissimum* L. was a principal oil and fibre source in the Old World and probably the earliest cultivated plant used for weaving clothes. Until recently, flax was extensively cultivated all over the area from the Atlantic coast of Europe in the west, to Russia and India in the east, and Ethiopia in the south (Durrant 1976). In antiquity, flax fibres (which are stronger than cotton or wool) were the principal vegetable fibre used for weaving textiles in Europe and western Asia. But from the industrial revolution onwards, flax was gradually replaced by cotton imports and more recently, it was almost completely replaced by cotton and synthetic fibres. The seed contains about 40 per cent oil and in peasant communities linseed was used as a source for edible oil and high-grade lighting oil.

Flax is an annual crop with characteristic slender, strong stems and rounded capsules which (in cultivated forms) do not dehisce but retain the oval, compressed, shining seed. The crop is diploid ($2n = 30$) and predominantly self-pollinated. Consequently, variation has been moulded in the from of numerous true breeding lines and aggregates of land races. Two specializations are apparent: (i) oil varieties which are relatively short (30–70 cm), branched, and usually bear large seed – they are grown for high yield of linseed; (ii) fibre varieties which are taller, sparsely branched, and usually produce small seed. Transitional forms cultivated for both oil and fibre occur as well.

Cooking oil is obtained from linseed by cold pressing. Drying oil (used for paints and varnishes) is produced by hot treatment of the seed before pressing (hot pressing).

The fibres for spinning are obtained from fibre-cell bundles running the length of the stem and forming a ring in the cortex. The stems are harvested before the maturation of the seed. Traditionally they were first dried, then immersed (or wetted) in water to allow the microbial decomposition ('retting') of the pectin connecting the fibres with other cells and tissues of the stem. After retting, the stems were dried and the fibres (averaging 4 cm in length) were separated by pounding ('breaking') and combing.

Flax is represented in archaeological excavations both by seed and, occasionally, capsules, and by remnants of stems or textiles. In the latter, fibres can be identified microscopically, if they are not carbonized.

Wild ancestry

Linum L. is a relatively large genus comprising some 200 species spread over the temperate, Mediterranean and steppe belts of the northern hemisphere. Cultivated flax, *Linum usitatissimum* L., is most nearly related to wild *L. bienne* Mill. (syn. *L. angustifolium* Huds.). These two flaxes have the same chromosome number ($2n = 30$), intercross readily, and are fully interfertile (Gill and Yermanos 1967). *L. bienne* (Fig. 34) with its characteristic strong branches, blue flowers, and dehiscent capsules is widely distributed over west Europe, the Mediterranean basin, North Africa, the Near East, Iran, and Caucasia (Map 12). Some wild forms are biennial or perennial, others are annual; all are predominantly self-pollinated. *L. bienne* grows mainly in wet places such as moist grassy areas, springs, seepage areas on rocky slopes, moist clay soils, and marshy lands. Occasionally it is also found as a weed in fields and at the edges of cultivation, including flax cultivation. In such places, hybridization between wild and cultivated flaxes occurs sporadically. On the basis of its close morphological and genetic affinities to the cultivated crop, *L. bienne* is now identified as the wild progenitor of *L. usitatissimum*. It is therefore more correct to refer to this wild flax as *L. usitatissimum* subsp. *bienne*. As in many other grain crops, the main changes under domestication were the shift to non-dehiscent capsules, the increase of seed size, and the selection for higher oil yield, or longer stems with a high amount of long fibre.

Archaeological evidence

The oldest linseed remains retrieved from excavation sites in the Near East and Europe have been assigned to wild *L. bienne*. They come from pre-farming (8000–7600 bc) Tell Mureybit, Syria (van Zeist 1970), and Early Neolithic (*c.* 7000 bc) Çayönü, Turkey (van Zeist 1972), 7500–6750 bc Ali Kosh, Iran (Helbaek 1969).

Fragments of a capsule from Pre-Pottery Neolithic B Jericho (Hopf 1983) may represent cultivated flax. Another indication of early flax cultivation comes from linseed remains recovered from 6250–5950 bc levels of Tell Ramad, Syria (Fig. 36). The calculated size of the seed from this site, corrected for charring shrinkage, ranges from 3.2 to 4.1 mm in length. This is already within the size class of the *L. usitatissimum* seed, the lower limit of which lies at 3.0 mm. It is therefore a good indication for flax cultivation under rain-dependent conditions before 6000 bc (van Zeist and Bakker-Heeres 1975a). In the Mesopotamian basin linseed sizes in 5500–5000 bc Tell Sabz (Helbaek 1969) and in Halafian 5000–4500 bc

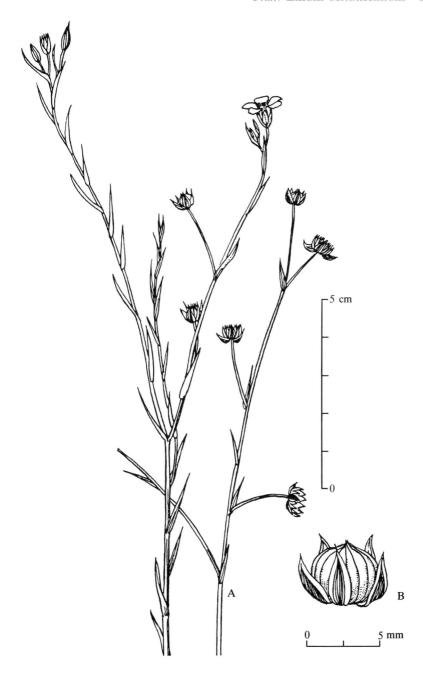

Fig. 34. Wild flax, *Linum usitatissimum* subsp. *bienne* (= *L. bienne*). A – Flowering and fruiting stem. B – Capsule. (M. Zohary 1972, plate 374.)

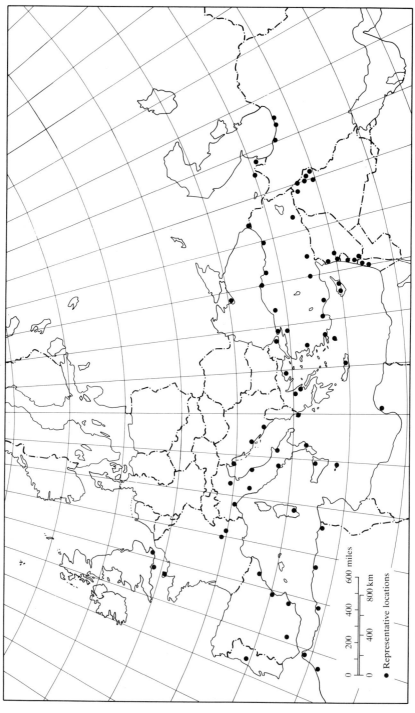

Map 12. Distribution of wild flax, *Linum usitatissimum* subsp. *bienne* (= *L. bienne*).

· Representative locations

0   200   400   600 miles

0   400   800 km

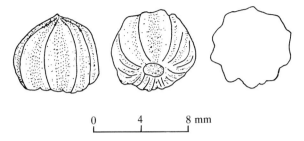

0   4   8 mm

Fig. 35. Carbonized capsule of flax, *Linum usitatissimum*. Bronze Age Jericho. (Hopf 1983.)

Arpachiya (Helbaek 1959a) are even bigger (4.7–4.8 mm long). As argued by Helbaek (1959a, 1985), these large seeds indicate advanced domestication and demonstrate that flax was part of the irrigation agriculture system that evolved in this region. Linseed also appears in several other Neolithic and Bronze Age sites in the Near East. A complete capsule was retrieved from Bronze Age Jericho (Fig. 35). Flax was also discovered, together with emmer wheat and barely in 5th millennium bc Merimde (Stemler 1980; M. Hopf, unpublished) and Fayum (Caton-Thompson and Gardner 1934; Wetterstrom 1984) in Egypt.

The spread of flax from the Near East 'nuclear area' into Europe is also well documented (Map 13). Linseed has been recovered from several early Neolithic sites in Thessaly, Greece (Kroll 1981a). Additional finds in this country are from *c.* 2400 bc Lerna (Hopf 1961b). In contrast, in central and western Europe, cultivated flax finds are from relatively late absolute dates,

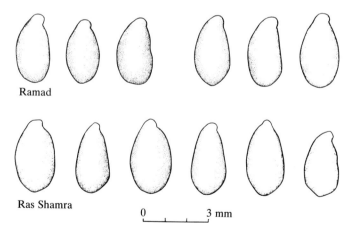

Ramad

Ras Shamra   0   3 mm

Fig. 36. Carbonized seed of flax, *Linum usitatissimum*. A – Neolithic Ramad, Syria. B – Neolithic Ras Shamra, Syria. (van Zeist and Bakker-Heeres 1975a.)

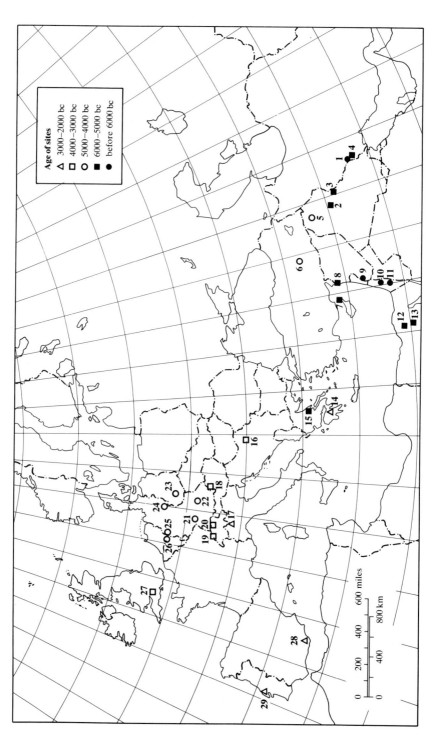

Map 13. The spread of flax to Europe, west Asia, and the Nile Valley. (Adapted from van Zeist 1980, fig. 3.)

but they belong nevertheless to the earliest food production cultures there, i.e. the Bandkeramik sites in the Aldenhovener Platte, Köln-Lindenthal, Hienheim and Heilbronn, Germany (Willerding, 1980; Table 4). Flax is common among plant remains of the Lake dwelling sites (Late Neolithic) in Switzerland such as Niederwil (van Zeist and Casparie 1974) or Zürich (Jacomet-Engel 1980). It is also known from Early Bronze Age northern Italy; that is, from Lagozza (Buschan 1895, p.241) and from Valeggio (Villaret-von Rochow 1958), and from several other sites in Switzerland, Germany, and Spain dated to this period. Much less is known about the spread of flax into Asia. Yet flax seeds were discovered in 3rd millennium bc Shahr-i Sokhta, east Iran (Costantini and Costantini-Biasini 1985).

Remains of flax textiles also appear early. The best examples come from drier parts of the Near East where, due to low humidity, woven material survived without carbonization. Pieces of exquisitely woven linen were recently discovered among Pre-Pottery Neolithic B remains in Nahal Hemar cave near the Dead Sea (Bar Yosef and Alon 1988; Schick 1988). They were radiocarbon-dated to the beginning of the 7th millennium bc. A single piece of linen was found in Neolithic (5th millennium bc) Fayum, Egypt (Caton-Thompson and Gardner 1934, p. 59). Linen fragments were retrieved from the Chalcolithic (*c*. 3400 bc) 'Cave of the Treasure' near the Dead Sea (Bar-Adon 1980). Flax textiles were common in the Old Middle, and New Kingdoms of Egypt, where linen was extensively used for wrapping mummies (Täckholm 1976; Germer 1985, p.100). Egyptian retting, spinning, and weaving of flax is beautifully recorded in 12th dynasty Beni Hasan grave paintings (Täckholm 1976).

In conclusion, the available archaeological evidence clearly suggests that flax belongs to the first group of crops that started agriculture in the Near East. Although this crop does not give us clear-cut signs to decide on the

---

Archaeological sites in Map 13:
1: Ali Kosh (Helbaek 1969). **2:** Tell es-Sawwan (Helbaek 1964b). **3:** Choga Mami, Mandali (Helbaek 1972). **4:** Tepe Sabz (Helbaek 1969). **5:** Arpachiyah (Helbaek 1959b). **6:** Girikihaciyan (van Zeist 1979–80). **7:** Andreas Kastros (van Zeist 1981). **8:** Ras Shamra (van Zeist and Bakker-Heeres 1986a). **9:** Ramad (van Zeist 1976). **10:** Jericho (Hopf, 1983). **11:** Nahal Hemar (Schick 1988). **12:** Merimde (Stemler 1980: M. Hopf, unpublished). **13:** Fayum (Caton-Thompson and Gardner 1934, p. 49). **14:** Lerna (Hopf 1951). **15:** Sesklo (Kroll 1981a). **16:** Gomolava (van Zeist 1975). **17:** Lagozza and Fiavè (Buschan 1895, p. 241; Villaret-von Rochow 1958). **18:** Mondsee (Neuweiler 1905). **19:** Twann, Bielersee (Piening 1981). **20:** Zürich (Jacomet-Engel 1980). **21:** Heilbronn (Willerding 1980, table 4). **22:** Hienheim (Bakels 1978). **23:** Eisenberg (Rothmaler and Natho 1957). **24:** Rosdorf near Göttingen (Wilerding 1980, table 4). **25:** Aldenhoven, including Langweiler, Lamersdorf, Bedburg-Garsdorf, and Rödingen (Knörzer, 1979; Willderding 1980, table 4.) **26:** Beck-Kerkeveld (Bakels 1978). **27:** Windmill Hill (Helbaek 1952c). **28:** El Argar (Buschan 1895; Hopf 1991b) and Almizaraque (Netolitzky, 1935; M. Hopf, unpublished data). **29:** Vila Nova de S. Pedro (do Paço 1954).

start of its domestication, the gradual increase in seed size and the use of linen indicate that flax cultivation was, very probably, already practised in the Near East before 6000 bc. The evidence from the living plants fully supports a Near East domestication. Wild *bienne* forms are widespread in the 'arc'.

Hemp: *Cannabis sativa*

Hemp, *Cannabis sativa* L., is a tall (2–3.6 m) dioecious, wind pollinated herb, and a member of a special small family Cannabinaceae. Hemp is cultivated for three main purposes: (i) for fibres obtained from the bast of the stems; (ii) for the seed that are used either for extraction of oil or as an animal feed; (iii) as a source of a psychotomimetic drug produced by the glandular hairs of the plant. Special cultivar groups have been developed for the different uses (Simmonds 1976). Fibre varieties are tall, succeed in both temperate and tropical climates, and contain negligible quantities of the drug. Hemp textiles are strong but coarse and in traditional communities they were used for cheap clothing, sacks, rags, or sails. Hemp oil is used mainly for technical purposes such as varnish or soap. The drug is obtained from special varieties (frequently referred to as the *indica* group) grown in hot climates. Only the tops of the female plants contain appreciable quantities of the *Cannabis* drug and are harvested for the preparation of marijuana or hashish.

Cultivated hemp is closely related to, and fully interfertile with, an extraordinarily variable aggregate of wild and weedy forms. Except for traits associated with the economic use, wild forms differ from the cultivars by relatively smaller achenes ('seed') which also possess adhering perianths and elongated bases. The crop is obviously a central Asiatic element, and temperate territories in this vast area such as the Caspian basin, parts of Afghanistan, and central Asia, or the Himalayas harbour spontaneous *C. sativa* plants which seem to be wild. Very probably populations which appear to grow wild are not fully primary but have been introgressed extensively with the cultivated varieties and weedy types which also abound in these areas (Small and Cronquist 1976). The picture of *Cannabis* is even further complicated by the marked tendency of this cross-pollinated crop to revert to wild. Naturalized derivatives and weedy races of *C. sativa* are now distributed not only in central Asia, but also over many other places in Asia, Europe, and America.

Archaeological records on the early establishment of hemp are not available so far. But this plant must have been taken into cultivation quite early – somewhere in temperate Asia. On the basis of linguistic and cultural evidence Li (1974) concluded that hemp was probably cultivated in China by at least 2500 bc and that it was the only fibre available to the ancient peoples of northern and north-eastern China.

The more recent expansion of hemp to the Near East, the Mediterranean basin and Europe is better documented. Remains of hemp fabrics are available from eighth-century bc Gordion, Anatolia. It also seems certain that the plant was known to the Sarmatians and Scythians who occupied the southern part of Russia between approximately 700 and 300 BC (Godwin 1967). At that time hemp became a well-known fibre crop in the Near East and Greece and spread to Italy and Sicily in about 100 BC. Its fibres were used among other things to make ropes and sails. The spread of *C. sativa* to central and north Europe from Roman times on is now documented by a wealth of macro-remains, palynological evidence and written sources (Dörfler 1990). The narcotic properties of *C. sativa* were recognized in India by 1000 bc, but apparently this knowledge did not reach the Mediterranean region before post-classical times.

Old World cottons: *Gossypium arboreum and G. herbaceum*

Two species of cotton, tree cotton, *Gossypium arboreum* L., and its close relative, short staple cotton, *G. herbaceum* L., were traditional fibre plants of the Old World. Both are diploids, have an almost identical chromosomal constitution ($2n = 26$; genomic designation AA) and bear relatively short staple lint less than 22 mm long (Lee 1984). These were the only *Gossypium* species grown in the warmer parts of Asia and Africa before the discovery of America. Since then the Old World cottons have been replaced by their American counterparts, i.e. tetraploid ($2n = 52$; genomic designation AADD) *Gossypium hirsutum* L. and *G. barbadense* L. which produce longer (24–35 mm) lint. Today *G. arboreum* and *G. herbaceum* survive as relic crops mainly in India, south-east Asia, and Africa. They account for less than 1 per cent of world cotton fibre production.

Carbonized seeds of cotton are reported (Costantini 1984) from the Neolithic (5th millennium bc) beds of Mehrgarh, Pakistan. However, these seeds are poorly preserved. It is doubtful whether they represent cultivated cotton or cotton as such. Definite indications of cotton cultivation and cotton use in the Indian subcontinent are available only from Harappan contexts (Hutchinson 1976). Fragments of cotton textiles and cotton strings have been found in Mohenjo-Daro, Pakistan in contexts now dated at about 1800 bc (Hutchinson 1976), as well as in contemporary Harappa and several other Pakistani and Indian sites (Vishnu-Mittre 1977). Tree cotton, *G. arboreum*, seems to have moved from India to the Near East in the 1st millennium bc. An early reference to its arrival comes from Assyria where we are informed that 'trees bearing wool' were introduced by Sennacherib in about 694 BC (Thompson 1949, p. 113). The establishment of *G. arboreum* (and also somewhat later of *G. herbaceum*) in the Near East and the east Mediterranean region during Hellenistic and Roman times, is well documented in Greek, Roman, and Jewish writings (Lenz 1859; Feliks 1983).

The wild *Gossypium* species of east Africa, south Arabia, and the Indian subcontinent are still insufficiently known. Consequently, the places of origin of cultivated *G. arboreum* and *G. herbaceum* are not yet firmly established. Spontaneous, wild-looking *arboreum* forms occur in India. They are unknown from Africa or anywhere else. If these Indian forms are genuinely wild (Hutchinson, 1976, regarded them as feral) the Indian subcontinent seems to be the place in which the tree cotton was taken into cultivation. The closely related *G. herbaceum* presents another story. Wild forms of this diploid cotton are widespread in east Africa. They do not occur in India. Hence the short staple cotton is very likely an African domesticate. An indication of such origin comes from 2500 bc Afyea, Egyptian Nubia, where cotton seed and lint hairs intermediate between those borne by wild forms and those produced by *herbaceum* cultigens, were discovered in goat coprolites (Chowdhury and Buth 1970, 1971). However, these finds are not accompanied by signs of cotton weaving in that area.

Poppy: *Papaver somniferum*

Poppy, *Papaver somniferum* L., is grown for two purposes (Duke 1973). First, it is famous as a source of opium which is obtained from the latex released by the plants after gashing their unripe capsules. It is also cultivated for its seeds which are rich in oil. The seeds are consumed as such or used to extract poppy oil. Because of this dual use two series of cultivars have evolved in *P. somniferum*. The opium forms are grouped in subsp. *somniferum* Corb., while the oil varieties are collectively known as subsp. *hortensis* (Hussenot) Corb. The oil is used both for eating and for industrial purposes. The opium is a powerful medicinal and narcotic element. Its effects, including pain killing, were already appreciated in antiquity.

Papaver somniferum is predominantly self-pollinated. Most cultivars are diploid and contain $2n = 22$ chromosomes. The crop is closely related to wild and weedy poppies which grow in the western part of the Mediterranean basin (Map 14). These were traditionally called *P. setigerum* DC.

The most conspicuous differences between the tame and the wild plants are the considerable increase in capsule size, and the retention of the seed in the capsule in poppies under domestication. Cultivated forms also show a tendency to increased seed size. Yet some wild forms bear fairly large seeds which considerably overlap the dimensions (and seed-coat reticulate sculpture, see Fig. 37) found in cultivated poppy seed. The seed size is therefore an unreliable trait for recognition of domestication in this crop (Fritsch 1979).

Wild *setigerum* poppies comprise both diploid ($2n = 22$) and tetraploid ($2n = 44$) chromosome types. The diploid *setigerum* forms were found to be fully interfertile with the somniferum cultivars (Hammer and Fritsch

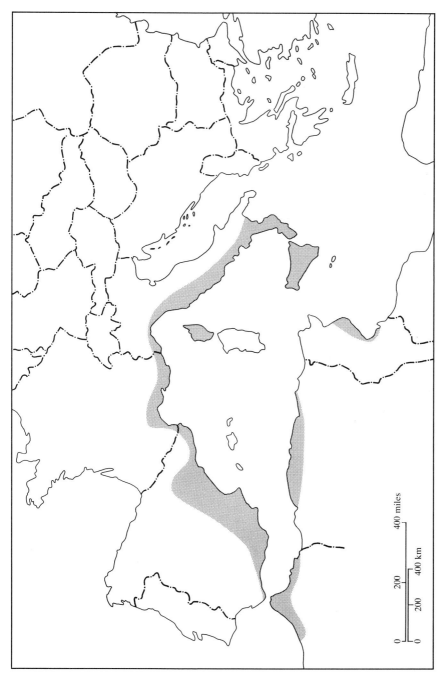

Map 14. Distribution of wild poppy, *Papaver somniferum* subsp. *setigerum*. (Based on Bakels 1982.)

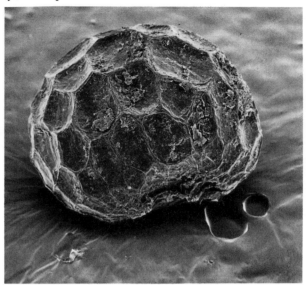

Fig. 37. The reticulate sculpture of the seed coat in the poppy, *Papaver somniferum*, × 55. (From Schoch *et al.* 1988.)

1977). They are therefore assumed to be the progenitor stock from which the cultivated poppy evolved. This necessitated also a taxonomic revision of the rank of *P. setigerum*. Most workers place it now as the wild race, subsp. *setigerum* (DC.) Corb., of the *P. somniferum* crop complex.

Few carbonized poppy seeds were found in the Danubian settlements of the Aldenhovener Platte, north-western Germany (Knörzer, 1971) and a few more in Zesławice, near Kraków, Poland (Giżbert 1960).

Numerous remains of charred seed, as well as occasional capsules are available from middle and late Neolithic central Europe (for review see Schultze-Motel 1979a): they come from middle Neolithic Mennevile, north France (Bakels 1984), from 4th millennium bc lake-shore settlements in Switzerland (Jacomet *et al.* 1991), contemporary sites in Germany, and late Neolithic Lagozza in north Italy. Four dessicated, beautifully preserved poppy capsules were reported by Neuweiler (1935) from Cueva de los Murciélago near Albuñal, Granada Province, Spain. The find comes from beds regarded to be Final Neolithic (c 2500 bc).

Significantly, poppies were not discovered in Neolithic sites in south-eastern Europe and in the Near East. They appear, however, in several Bronze Age sites in Greece, Bulgaria, and the former Yugoslavia (Kroll 1991). As stressed by van Zeist (1980) and by Bakels (1982) this fact, combined with the distribution area of wild *setigerum* poppies, strongly suggests west Mediterranean domestication. In other words, *P. somniferum* does not belong to the primary 'first circle' Near East crops which started food production in Europe. It is a representative of the 'second circle' domesticants, i.e. crops that were added to the original assemblage already outside the Near East core area.

False flax: *Camelina sativa*

False flax, *Camelina sativa* (L.) Crantz., of the mustard family Cruciferae, is a relic oil plant rapidly disappearing from cultivation. Yet until the 1940s, *Camelina* was an important oil crop in eastern and central Europe. *Camelina* seed, with their characteristic protruding embryos, appear repeatedly in European archaeological contexts.

The crop, subsp. *sativa* (Mill.) E. Schmid, is closely related to, and interfertile with, a variable aggregate of wild and weedy forms distributed over Europe and south-west Asia (for review, see Markgraf 1975). Truly wild, late-flowering forms, now recognized as subsp. *microcarpa* (Andrz.) E. Schmid but formerly referred to as *C. microcarpa* Andrz., grow in the east European steppes and adjacent Asiatic territories. They have characteristic small fruits. Closely related to them are early flowering hairy forms, known as subsp. *pilosa* (DC.) E. Schmid which thrive in fields of winter cereals almost all over Europe. These are obviously recently evolved weeds, which spread from east Europe westward, over cultivated lands. An additional distinct *Camelina* weedy race, namey subsp. *alyssum* (Mill.) E. Schmid (= *C.sativa* var. *linicola* Prusch.) evolved in association with flax cultivation in Europe. These *linicola* forms, with their long erect stems and hard fruits, serve as a well-documented example of evolution of weed mimicry to a specific cultivated crop. The cultivated varieties of *C. sativa* differ from the wild and weedy forms by their non-dehiscent, large and pyriform fruits and by their bigger (1.5–2.0 mm long) seed, which contain an appreciable amount (27–31 per cent) of edible oil.

Camelina seeds (as well as occasional pods) appear repeatedly in central and eastern European archaeological contexts. However, they start late. The earliest finds come from 3rd and 2nd millennia bc beds. Large samples and/or pure samples, i.e. material that can be regarded as the cultivated crop, appear even later.

In central Europe the oldest remains come from Final Neolithic beds (*c.* 2000 bc) of Auvernier, Switzerland; and these are followed by several Bronze Age (1800–1200 bc) finds from Poland, Hungary, Germany, and north Italy (Schultze-Motel 1979b; Wasylikowa *et al.* 1991). Iron Age finds are more numerous at this stage. *Camelina* remains become particularly common in the coastal areas of the Baltic and the North Sea. At least some of these finds (particularly the large samples with bigger seeds) seem to represent *Camelina* cultivars.

Further south-east, a single seed of *Camelina* was found in Chalcolithic (*c.* 3000 bc) Pefkakia, Thessaly (Kroll 1991) and many more in *c.* 2200 bc Sucidava-Celei, Rumania (Wasylikowa *et al.* 1991). *Camelina* seeds were also retrieved from Early Bronze Age Demircihüyük in northwest Anatolia (Schlichtherle 1977/78). They reappear in several Late Bronze Age sites in Greece, Bulgaria, and the former Yugoslavia (Kroll 1991); and also in Late Bronze Age Hadidi on the Syrian – Turkish border (Miller 1991). In the latter site their concentration again suggests cultivation.

All in all, the combined evidence from the living plants and from archaeology indicates that *C. sativa* is a secondary crop. Very probably this crucifer entered agriculture first by evolving weedy races that infested flax and cereal cultivation. Only later the weed was picked up as an oil crop.

Other cruciferous oil crops

The cultivation of several other members of the mustard family (Cruciferae) is amply recorded in classical times. They were appreciated as oil or mustard sources (extracted from their seed) and/or for their vegetable parts. Foremost among these crucifers were the radish *Raphanus sativus* L., the various mustards, i.e. the white mustard *Sinapis alba* L., black mustard *Brassica nigra* (L.) Koch, brown mustard *B. juncea* (L.) Czern. and the turnip *B. campestris* L. (Syn. *B. rapa* L.). All these crops have wild forms distributed over west Asia and Europe. They include variable, aggressive races of weeds which infest agricultural land far beyond these territories. It is very likely that seed of wild forms (*Brassica, Sinapis*, and also *Camelina* and *Descurainia*) were collected as oil sources in Switzerland and Germany in late Neolithic times (Schlichtherle 1981).

Because these crucifers were already well-established oil crops in Hellenistic and Roman times, it is safe to assume that they were taken into cultivation earlier. Yet up to now there are almost no archaeological records available for any of these crops. (For the scanty information present, see Renfrew 1973, pp. 166–7, and the annual reviews of Schultze-Motel 1968–90.) Suggestions as to the origin of these plants are necessarily based on linguistic considerations.

A single rich find of cruciferous seed is, however, described from 3000 bc Khafajah, Iraq. Here numerous carbonized seeds were discovered in several locations at the Temple Oval (Bedigian and Harlan 1986). They were examined in the 1930s by E. Schiemann who concluded that they belonged to *Brassica* or *Sinapis*. (It is very difficult to distinguish between these two genera in charred material.) Thus there is at least one archaeobotanical indication that crucifers might have been used in Mesopotamia in relatively early times.

Sesame: *Sesamum indicum*

Sesame, *Sesamum indicum* L. (syn. *S. orientale* L.) is a traditional warm season crop in south-west Asia and the Mediterranean basin. It is highly appreciated for its oil which keeps fresh for a long time without turning rancid. Sesame was extensively cultivated in the Graeco-Roman world (Lenz 1859; Gallant 1985), and at that time apparently more for its edible seed than for its pure oil. As stressed by Gallant (1985), Theophrastos categorizes sesame, along with the millets, as one of the main summer crops of his time. But despite such wide Hellenistic and Roman use, sesame does not belong to the Near East crop assemblage; it probably arrived from further east and rather late. Undisputed remains are available in this region only from the 1st millennium bc. Claims of earlier introduction into Mesopotamia are based on references to 'Šmaššamū' (Akkadian) and 'Še-giš-ì' (Sumerian) in Babylonian tablets (Postgate 1985). But most workers now contend that these terms denote 'oil plant' in general and do not provide clear evidence for the cultivation of *S. indicum* in 2nd millennium bc in Mesopotamia. Seasonal considerations are not clear either.

The botanical evidence also supports a late introduction. Wild *Sesamum* species are totally absent in the Near East and the Mediterranean basin. The genus is restricted to Africa south of the Sahara (numerous wild species) and to the Indian subcontinent (few). A group of wild and weedy forms native to India and described as *S. orientale* L. var. *malabaricum* Nar. shows close morphological, genetic, and phytochemical affinities to the crop. Bedigian and Harlan (1986) recently proposed them to be the progenitor of the cultivated *S. indicum*. If indeed the wild and weedy *malabaricum* forms represent the wild ancestor, domestication of sesame should have started in India. Indeed, the oldest record of *S. indicum* cultivation comes from Harappa in the Indus Valley (for review see Bedigian 1985). No precise date has been established for this record; but most authors place it between 2250 and 1750 bc.

The earliest remains of sesame seeds in the Near East come from Armenia. Four large jars containing carbonized sesame seed were excavated in Karmir Blur on the outskirts of Yerevan. They were dated between 900 and 600 bc. The site also contains elaborate installations for the extraction of oil from the seed (Bedigian 1985, pp. 168–9). Another find of the same time comes also from the Urartu Kingdom (Bastam, Van district, east Turkey; (Hopf and Willerding 1989). In addition, some 200 sesame seeds were uncovered in Iron Age beds (*c.* 800 bc) at Deir Alla, Jordan (Neef 1989). Except for these finds no other sesame remains have been recovered from pre-classical Near East sites, although it should be no problem to identify *S. indicum* seed, if it occurs in examined plant remains.

It is therefore clear that *S. indicum* arrived in the Near East from the east. It was apparently taken into cultivation in the Indian subcontinent. It is still undecided when the crop was domesticated and how early it was introduced into Mesopotamia and other parts of the Near East.

5

Fruit trees and nuts

Fruit trees constitute an important element of food production in the countries bordering the Mediterranean Sea. Their long-standing economic importance is amply reflected in classical traditions (Stager 1985). Five of the Biblical 'seven species' are fruit trees. Olive oil, wine, dry raisins, dates, and figs were, and still are, staple agricultural products in the Near East and the Mediterranean basin.

Like the Neolithic grain crops, the first fruit trees seem to have been brought into cultivation in the Near East 'core area' (Zohary and Spiegel-Roy 1975). Yet horticulture seems to have started relatively late in the history of food production in the Old World. The first definite signs of fruit-tree cultivation in the Near East appear in Chalcolithic contexts (4th millennium bc), i.e. several millennia after the firm establishment of grain agriculture. In the Bronze Age, olives, grapes, and figs emerge as important additions to the cereals and the pulses throughout the eastern Mediterranean basin, including the Aegean belt; and dates are cultivated in the warmer southern fringes.

Horticulture is a technology very different from grain agriculture. Cereals and pulses are annual 'short investment' crops. Their grains are harvested several months after sowing. The grain cultivator is able to move from place to place after the harvest and practise shifting farming. In contrast, fruit trees are perennials, with orchards starting to bear fruit 3–8 years after planting and attaining full productivity several years later. Horticulture therefore indicates a fully sedentary way of life.

Genetically, domestication of fruit trees means first of all changing the reproductive biology of the plants involved (Zohary and Spiegel-Roy 1975), by shifting from sexual reproduction (in the wild) to vegetative propagation (under cultivation). As a rule, cultivated varieties of fruit trees are maintained vegetatively by cuttings, rooting of twigs, suckers, or by the more sophisticated technique of grafting, and they are seldom raised from seed. This is in sharp contrast with their wild relatives which reproduce from seed. In other words, wild populations maintain themselves through sexual reproduction and as a rule are distinctly allogamous. Cross-pollination is brought about either by self-incompatibility or by dioecy (separate male and female individuals). Spontaneous populations manifest wide variability and maintain high levels of heterozygosity. Consequently, seedlings raised from any mother tree segregate widely in numerous traits, including the size, shape, and palatability of the fruits.

In the hands of the grower, vegetative propagation has been a powerful device to prevent genetic segregation and to 'fix' desired types. By discarding sexual reproduction and inventing clonal propagation the farmer was able first to select, in one act, exceptional individuals with desirable fruit traits from among large numbers of variable inferior trees, and second to duplicate the chosen types to obtain genetically identical saplings. In the case of the fruit trees this is no small achievement. Because the plants are cross-pollinated and widely heterozygous, most progeny obtained from seed (even progeny derived from the best varieties) are economically worthless. The change from seed planting to vegetative propagation has been the practical solution to assure a dependable supply of desired genotypes. In most fruit trees this invention made cultivation possible. Only few have been traditionally maintained also by seed planting (e.g. almond, walnut).

Plant remains retrieved from archaeological excavations indicate that the olive, grape vine, fig, date palm, pomegranate, and sycamore fig were the first fruit trees cultivated in the Old World. Significantly, for all these 'first wave' fruits the cultivated clones can be multiplied simply by cuttings (in the grape, fig, sycamore fig, and pomegranate), rooting of basal knobs (in the olive) or by transplanting offshoots (in the date). The early cultivators did not have to resort, in these crops, to sophisticated techniques of vegetative propagation such as grafting. Wild olives, grapes, figs, dates, and pomegranates were thus 'preadapted' for domestication since they lent themselves easily to vegetative manipulation. This trait was probably decisive for their success in becoming the first domesticated fruit trees in the Near East.

Several other fruit trees, such as the apple, pear, plum, and cherry seem to have been taken into cultivation much later. Definite evidence for their cultivation appears only in the 1st millennium bc and their extensive incorporation into horticulture seems to have taken place only in Greek and Roman times. This is in spite of the fact that the wild relatives of these fruit trees are richly represented in west Asia and in Europe, and fruits of *wild* forms were frequently collected by Neolithic and Bronze Age farmers.

A plausible explanation for the late appearance of these 'second wave' fruit domesticates is that they do not lend themselves to simple vegetative propagation. Their culture is based almost entirely on grafting. The Greeks and the Romans were already familiar with this art and we have ample documentation that in classical times apples and pears were maintained by this sophisticated method of vegetative reproduction (for review, see White 1970, pp. 248, 257–8). When and where detached scion grafting was invented is not yet clear. But obviously the introduction of this advanced propagation method made possible the domestication of a new range of fruit trees. Very probably the initiation of grafting was outside the area of Mediterranean horticulture and was introduced into this region from the east. The earliest documentation on grafting comes from China in connection with citrus fruit cultivation (Cooper and Chapot 1977, p. 11).

The adoption of clonal cultivation means that most fruit trees, in the 5 or 6 millennia since their introduction into cultivation, have undergone very few sexual cycles. Cultivated clones persisted for hundreds or even thousands of years. From the standpoint of evolution under domestication this means a severe restriction on selection. In other words, selection could only have operated during a limited number of generations, and we have to expect that the cultivars have not diverged considerably from their progenitors' gene-pools. This is in sharp contrast to the grain plants, in which selection could have operated during thousands of generations. Indeed, the cultivated varieties of fruit trees can be regarded as exceptional, highly heterozygous individuals of their biological species, clones that excel primarily in fruit size and fruit quality. The absence of profound genetic changes in the fruit trees under domestication is also apparent in their ecology. The climatic requirements of the cultivars closely resemble those of their wild relatives. Unlike cereals and pulses, the fruits have not been pushed much beyond the climatic requirements of their wild ancestors.

As already stated, most fruit trees under cultivation are derived from allogamic wild progenitors in which cross-pollination is maintained either by self-incompatibility or by dioecy. Because of this background, the shift from sexual reproduction to planting of vegetatively propagated clones introduced serious limitations on fruiting. Planting of a single self-incompatible clone, or alternatively female clones, would not bring about fruit set. Several agronomic devices assuring fruit set in the orchard have been empirically adopted. They were also accompanied by unconscious selection for several types of mutations that resolved the restrictions set by self-incompatibility and sex determination.

In hermaphroditic, self-incompatible species such as the olive, apple, or pear, the early planters very likely realized that to obtain satisfactory fruit set it was necessary to plant together two or more synchronously flowering clones. The traditional cultivation of such fruit trees is based on mixed planting, a practice which brings about pollination between different genotypes. Two additional solutions are the outcome of unconscious selection. In the peach, apricot, sour cherry, European plum, as well as in several cultivars of almonds or olives we find mutations that caused the breakdown of self-incompatibility, or at least rendered this system 'leaky', so that self-pollination also brings about considerable fruit set in such clones. In other self-incompatible stocks pollination has been dispensed with altogether by incorporation of mutations conferring parthenocarpy (fruit development without fertilization and without seed development). Several clones of cultivated pear show this adaptation.

In dioecious species, fruit set is safeguarded in parallel ways. In the pistachio 'mixed planting' is employed, and some male individuals are planted together with the female clones. In date palms and Smyrna type figs, natural pollination is replaced by artificial pollination. Also in dioecious fruit trees the early planters must have selected unconsciously two types of mutations that rendered cross-pollination unnecessary. A genetic change from dioecy (in the wild) to hermaphroditism (in cultivation) evolved in the grape. A replacement of pollination by parthenocarpy occurred in clones of fig, sycamore fig, and in the Corinth type grapes.

Olive: *Olea europaea*

The olive, *Olea europaea* L., is the most prominent, and economically perhaps the most important classical fruit tree of the Mediterranean basin. Together with the grape vine, fig, and date, it comprises the oldest group of plants that founded horticulture in the Old World (Zohary and Spiegel-Roy 1975; Stager 1985). Since the Bronze Age, the wealth of many Mediterranean peoples centred around the cultivation of olives which provided valuable storable oil as well as edible fruits. Olive oil was used in eating and cooking, as well as for ointment and lighting. Because of its excellent keeping qualities, it served as a principal article of commerce. The whole fruits were also preserved and eaten. Bread and olives were – and still are – a staple diet in peasant communities throughout the Mediterranean basin.

Olives grow only in typical Mediterranean climates and the cultivated varieties and closely related wild *oleaster* forms are considered as reliable indicators of Mediterranean environments. The olive is a relatively slow-growing fruit tree. Production starts 5–6 years after planting. Olives under cultivation manifest considerable variation in the size, shape, and oil content of their fruits. Hundreds of distinct varieties are recognized and different parts of the Mediterranean basin are frequently characterized by specific local forms. The cultivated olives can be roughly subdivided into two main types: (a) oil varieties, the ripe fruits of which contain at least 20 per cent, and usually 25–30 per cent oil, and (b) table olives: less oily forms used for the preservation of whole fruits. There are also dual purpose varieties. Carbonized stones and charred wood of pruned twigs, used as firewood, constitute the bulk of olive remains in archaeological excavations.

Wild ancestry

The cultivated olive, *O. europaea* L., shows close affinities to a group of wild and 'weedy' olives distributed over the Mediterranean basin and traditionally called 'oleaster' olives. These wild forms are fully interfertile with the cultivated varieties and are interconnected with them by sporadic spontaneous hybridization and by feral types (Zohary and Spiegel-Roy 1975). Oleaster olives and the cultivated clones have more or less the same climatic adaptation. Previously, wild *oleaster* olives were treated by some botanists as an independent species: *O. oleaster* Hoffm. & Link. But because of its close morphological and genetic affinities to the cultivated fruit tree, most researchers dealing with Mediterranean plants today regard *oleaster* as the wild race of the cultivated crop, and place it within *O. europaea* L., either as a subspecies (subsp. *oleaster* (Hoffm. & Link) Hegi), or as a variety (var. *sylvestris* (Mill.) Lehr. = var. *oleaster* (Hoffm. & Link) DC.).

Oleaster olives differ from the cultivated clones mainly by their smaller fruits and usually also by spinescent juvenile branches. The fruits have a less fleshy mesocarp and contain less oil, although the stones are not considerably smaller than in cultivated olives. Wild olives are self-incompatible and reproduce entirely from seed, and as with many other trees maintaining such a genetic system, *oleaster* populations often show a wide range of variation. Under domestication the reproductive biology of the tree has changed; cultivated varieties are maintained by vegetative propagation and are, in fact, clones. Vegetative propagation depends primarily on the utilization of knobs (uovuli) that develop at the base of the trunk, and root easily when cut off. Trees may also be propagated by truncheons, by cuttings, and by grafting.

In *Olea*, as with numerous other fruit trees, the shift to vegetative propagation is the cultivator's countermeasure to the problems of wide genetic segregation which is characteristic of sexual reproduction in cross-pollinated plants. Cultivated olive clones are usually extremely heterozygous. When their seed are planted, the ensuing progeny segregate freely. In fact, most seedlings raised from a cultivar resemble wild forms in their morphology and are useless in terms of fruit quality. Consequently propagation from seed is impractical in oleoculture. In order to 'fix' useful genotypes, the grower has to resort to clonal propagation. Seedlings can only be used as a variable raw material for selection of new clones. Today such screening of seedlings is performed in some olive-breeding programmes. But spontaneous or subspontaneous seedlings have doubtless accompanied olive growing from its very start, and rare individuals showing superior qualities may have caught the attention of early cultivators and were then picked up as new clones.

Over large areas in the Mediterranean basin *oleasters* thrive as a constituent of *maquis* and *garrigue* formations (Map 15). In addition, these small trees frequently colonize secondary habitats such as the edges of cultivation or abandoned orchards. In such places, particularly in areas of olive plantations, spontaneous olives seem frequently to be 'feral', i.e. secondary sexual derivatives of the vegetatively propagated cultivated

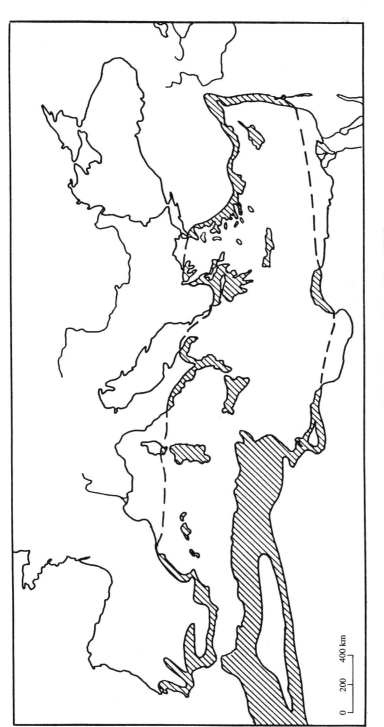

Map 15. Distribution of wild olive, *Olea europaea* subsp. *oleaster*. (After Zohary and Spiegel-Roy 1975.)

clones or products of hybridization between the cultivated clones and adjacent wild *oleaster* plants.

The Mediterranean wild olives are associated with olive cultivation in another way: they were often valued as stock material for grafting. *Oleaster* suckers or knobs were dug out in the wild and planted in the orchards as hardy stock material for the grafting of cultivated scions. In some localities, for example, in western Turkey, one still encounters another old tradition, namely the grafting of wild olives *in situ*, i.e. in their natural non-arable, *maquis* or *garrigue* habitats. The grafted or partly grafted wild olives are protected by the peasants while other trees and shrubs are cut for firewood and further suppressed by intensive grazing.

Olea europaea is the only Mediterranean representative of the genus *Olea* L., which otherwise includes some 35–40 species distributed over tropical and southern Africa (the main centre), south Asia, as well as eastern Australia, New Caledonia, and New Zealand. Both the cultivated olive and the Mediterranean *oleaster* are closely related to (and probably interfertile with) several non-Mediterranean wild olives. Most widespread among the latter are east African, south Arabian, south Iranian and Afghan wild olive forms frequently referred to as *O. africana* Mill. or *O. chrysophylla* Lam. (in Africa) and *O. ferruginea* Royle (in Asia). The morphological differences between these more tropical wild olives and their Mediterranean counterparts are relatively small. Consequently, Green and Wickens (1989) regarded them only as an additional subspecies of the European olive and named them *O. europaea* L. subsp. *cuspidata* (Wall.) Ciferri. However, these south Asian and east African olives are geographically fully separated from their Mediterranean relatives by wide discontinuities. They are also adapted to totally different environments. Such discontinuities fully justify their ranking as independent species.

Other non-Mediterranean wild-types closely related to the crop are *O. laperrinei* Batt. & Trab. and *O. europaea* L. subsp. *cerasiformis* (Webb & Berth.) Kunkel & Sunding (Syn. *O. europaea* L. var. *maderiensis* Hart.). The first is a Saharo-Montane relic that bridges the Mediterranean forms with their African savanna relatives. It is confined to few inner mountains in the Sahara desert and does not come close to the Mediterranean forms, except perhaps in the southern Atlas Mountains. The second is the wild olive of the Macronesian Islands. (For details on the distribution of all these non-Mediterranean wild olives consult maps 1 and 2 in Green and Wickens 1989).

To sum up, the evidence from the living plants shows that the cultivated olive is closely related to the Mediterranean *oleaster* wild forms; and these should be regarded as the wild stock from which the cultivated fruit tree had been derived. The Mediterranean olives (both wild and cultivated) are also reproductively well isolated from their non-Mediterranean wild relatives; the latter had nothing to do with olive domestication.

Archaeological evidence

It is likely that olives were collected from the wild long before their cultivation. The morphological similarity between stones of wild and cultivated
material renders the distinction between collection and cultivation in early
sites impossible where *oleaster* olives occur in masses in the same areas.
The few olive stones found in Natufian and Early Neolithic Nahal Oren,
Mt. Carmel, Israel (Noy *et al.* 1973) seem to represent a collection from the
wild. The numerous remains found in the 5th millennium bc Dhali Agridhi,
Cyprus (Stewart 1974) and in contemporary submerged Neolithic sites off
Mt. Carmel (Galili *et al.* 1989) were apparently of similar nature. However,
they could also represent an early stage of domestication.

Definite signs of olive cultivation come from Chalcolithic Palestine.
Numerous well-preserved carbonized olive stones (see Fig. 38), as well
as charred olive wood, were discovered in Chalcolithic (3700–3500 bc)
Tuleilat Ghassul, north of the Dead Sea (Zohary and Spiegel-Roy 1975;
Neef 1990) together with cereal grains, dates, and pulses. This site lies far
outside the natural range of wild olives, and no *oleaster* olives occur today
in the lower Jordan Valley or on the adjacent escarpments. The region is
too dry for this plant and this was probably true also in the Chalcolithic.
In Israel and Jordan, the areas nearest to the lower Jordan Valley that
support wild olives are the western flanks of the Judean Hills and Mt.
Carmel. This indicates that the Tuleilat Ghassul olives may have been
products of cultivation. They were probably raised under irrigation in a
similar manner in which olives are grown today in the Jordan Valley.

The finds from Tuleilat Ghassul are supplemented by stones and charcoal
remains from (a) three other Chalcolithic sites in the Jordan Valley, namely
Tell Saf (Gophna and Kislev 1979), and Tell Shuna North and Tell Abu
Hamid (Neef 1990) which also contain masses of fragments of crushed
stones, i.e. waste of olive pressing; and (b) contemporary sites in the
Golan Heights (Epstein 1978). Some centuries later, olive remains abound
in Early Bronze Age (2900–2700 bc) Arad (Hopf 1978a), Bab edh-Drah
(McCreery 1979) and Yarmuth (Ben-Tor 1975). Carbonized stones as well
as charred olive wood are available also from Early and Middle Bronze Age
Lachish (Helbaek 1958). Outside Palestine, early finds of olives have been
few so far. Olive remains were found in 3rd millennium bc Tell Soukas,
Syria (Helbaek 1962). Few are also present in several Early Minoan sites
such as Myrtos (Renfrew 1972) and Knossos (Evans 1928) in Crete. In the
Middle and Late Bronze Age, olive cultivation and olive-oil production
seem already to have been well-established throughout the countries
bordering the east shore of the Mediterranean Sea (Stager 1985) and in
Late Bronze Age also in Crete and mainland Greece (Boardman 1976;
Hansen 1988). The successful establishment and large scale utilization of
Olea is indicated also by the increase, from Early Bronze Age on, of

Fig. 38. Carbonized stones of olives, *Olea europaea*. Chalcolithic Tuleilat Ghassul, Jordan. (Zohary and Spiegel-Roy 1975.)

olive pollen grains in cores obtained from the Sea of Galilee (Baruch 1990) as well as by the appearance of numerous presses, and olive oil vessels and depiction of olives in Bronze Age art. Apparently the olive did not play a major role in Egypt. On the contrary, the export of olive oil from Palestine to Egypt is well documented in the Bronze Age (Stager 1985). Olive cultivation was probably introduced into the west Mediterranean basin in the early part of the 1st millennium bc by the Phoenician and Greek colonists (Boardman 1976).

In conclusion, in the Mediterranean basin olives constitute a complex of wild forms, weedy types, and cultivated clones. All are genetically loosely interconnected and all show similar climatic and edaphic preferences. This information from the living plants corresponds well with the available archaeological evidence. The earliest archaeological records of olive cultivation come from the lands bordering the eastern shores of the Mediterranean Sea, i.e. a territory where genuinely wild *oleaster* olives thrive today. Hence, on the basis of the combined evidence, one is led to the conclusion that the olive was probably first brought into cultivation in the Levant.

Grape vine: *Vitis vinifera*

Grape vine, *Vitis vinifera* L., is one of the classical fruits of the Old World. Together with the olive, fig, and date palm, it comprises the oldest group of fruit trees around which horticulture was developed in the Mediterranean basin (Zohary and Spiegel-Roy 1975). Since Early Bronze Age, grapes have contributed significantly to food production in this area, providing fresh fruits rich in sugar (the berries contain 15–25 per cent sugar), easily storable dried raisins, and juice for fermentation of wine. The latter became an important trade element in the countries around the Mediterranean Sea.

The grape vine thrives in Mediterranean-type environments, but it can tolerate cooler and more humid conditions than the olive. Thus viticulture extends northward beyond the Mediterranean basin and succeeds in areas with a relatively mild climate in western and central Europe and in western Asia. *Vitis* is a perennial climber that has to be pruned yearly in order to remain confined to a manageable size, and for regulation of fruit production. Under cultivation, grape vines are propagated vegetatively by rooting winter dormant twigs or by grafting. As in most other fruit trees, viticulture is based on the 'fixation' and maintenance of vegetative clones. Almost all cultivars bear hermaphrodite flowers and set fruit by self-pollination. Traditional Old World viticulture was based on thousands of distinct clones (Einset and Pratt 1975; Olmo 1976). These vary widely in their habit, in climate and soil preferences, and in the shape, size, colour, and sweetness of their fruits. Juicy, small-berried varieties with a rather acid taste are commonly used for wine production, especially in Europe. Types with sweet, large fruits prevail as table grapes. Some clones, such as the traditional Black Corinth and Sultanina, bear small, seedless berries appreciated in raisin production. Grape vine is a relatively fast growing crop. As a rule production starts 3 years after planting.

Carbonized pips and charred wood constitute the bulk of grape-vine remains in archaeological digs, although some sites also yielded parched whole fruits (containing two to four pips).

Wild ancestry

The cultivated grape *V. vinifera* L., is closely related to an aggregate of wild and feral vine forms (Fig. 39a) distributed over Europe and western Asia. Formerly, botanists regarded these wild grapes as an independent species: *V. sylvestris* C.C. Gmelin. But since these wild forms show a close morphological similarity to and are interconnected with the cultivated grape, most botanists regard the wild *sylvestris* grape vines today as the wild race of the cultivated crop. They place them as subspecies *sylvestris* (C.C. Gmelin) Berger within the *V. vinifera* crop complex and consider them as the stock from which the cultivated grape vine was derived (see, for example, Webb 1968).

Sylvestris grapes are widely distributed from the Atlantic coast to Tadzhikistan and the western Himalayas (Map 16). They are primarily forest climbers and seem to be native in the humid and climatically mild forest area south of the Caspian Sea and along the southern coast of the Black Sea. *Sylvestris* vines also abound in the cooler mesic northern fringes of the Mediterranean sclerophyllous vegetation belt from Turkey and Crimea, through Greece and Yugoslavia, to Italy, France, Spain, and north-west Africa. Along the Rhine and the Danube, wild grapes penetrated deeply into central Europe. Scattered colonies of wild grapes occur also in more xeric areas and in less woody places in the Near East arc. But here *sylvestris* vines are mostly confined to gorges and to the vicinity of springs and streams.

The boundary between the cultivated grape vine clones and the wild forms is blurred by the presence of escapees and secondary derivatives of hybridization. Spontaneous crossing between wild plants and cultivars has been found repeatedly where *sylvestris* vines grow in close proximity to vineyards; F_1 hybrids are fully fertile; so in *V. vinifera* we are faced in the Mediterranean basin with a variable complex of wild forms (growing in primary habitats), escapees and seed-propagated weedy types (which occur mainly in disturbed surroundings), and cultivated clones. The picture of the pre-agriculture distribution of the wild vine has probably been blurred by 'weedy' forms occupying secondary habitats. However, there can be little doubt that *sylvestris* vines are indigenous in southern Europe, the Near East, and the southern Caspian belt.

Sylvestris grapes differ from the cultivated varieties by their relatively small (Fig. 39a) and usually acid berries which are, however, quite suitable for the preparation of wine. Wild grapes may also be recognized by somewhat more globular pips, generally with stalks or 'beaks' constricted at the

0   20   40 mm

Fig. 39a. Fruiting wild grape vine *Vitis vinifera* subsp. *sylvestris*. (Zaprjagaeva 1964, plate 283.)

attachment to the main body of the pip (Fig. 39b). But these morphological differences are not very consistent and they cannot be regarded as fully safe diagnostic traits for distinguishing between wild and cultivated *Vitis* in archaeological remains.

Domestication has also brought about considerable changes in the reproductive biology of the plant. The first obvious change is the shift from the wild type sexual reproduction to the vegetative propagation of clones (by cuttings or by grafting). Just as in the olive, this is the cultivator's way of overcoming segregation in the seedlings and achieving 'fixation' of desired types. The second conspicuous development is the breakdown of the wild system of sex determination. *Sylvestris* grapes are dioecious and their populations contain an equal proportion of male and female individuals. Fruit setting in the wild depends on cross-pollination. In contrast, nearly all cultivated varieties are self-fertile: the flowers contain both pistils and

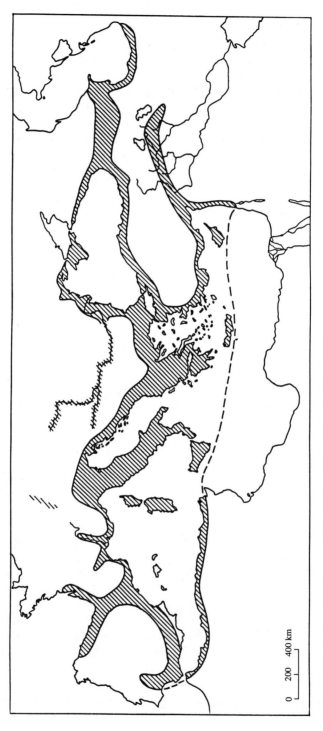

Map 16. Distribution of wild grape, *Vitis vinifera* subsp. *sylvestris*. (After Zohary and Spiegel-Roy 1975.) Note that toward the east the wild grape extends beyond the boundaries of this map and reappears in a few places in Turkmenistan and Tadzhikistan.

anthers and the hermaphroditic condition ensures self-pollination and fruit setting without the need of male pollen donors. Sex determination in wild *sylvestris* plants is governed by a single gene (Olmo 1976). Female individuals are homogametic, i.e. carry a homozygotic recessive genotype $Su^m Su^m$ that suppresses the development of anthers. Males are heterozygous for a dominant female suppressing Su^F allele and have a $Su^F Su^m$ genotype. The change under domestication to hermaphrodism was attained by a shift to Su^+ allele which is dominant over Su^m and brings about the development of both pistil and anthers in the flower. Many cultivated clones are still heterozygous and show a $Su^+ Su^m$ constitution; others have a $Su^+ Su^+$ genotype.

Vitis vinifera is the sole Mediterranean representative of the genus *Vitis*. This is a rather large genus comprising several dozen species (De Lattin 1939; Levadaux 1956; Mullins *et al.* 1992, p. 19). About two-thirds of them are centred in North America and one-third is distributed over east Asia. All members of *Vitis* are perennial woody climbers with coiled tendrils. All known *Vitis* species, excluding the *Muscademia* group which is now regarded as an independent genus, have $2n = 38$ chromosomes and can be easily crossed experimentally and their F_1 hybrids are vigorous and fertile. In nature, the principal types are in most cases separated from one another by geographical and ecological barriers. In some areas, this isolation is

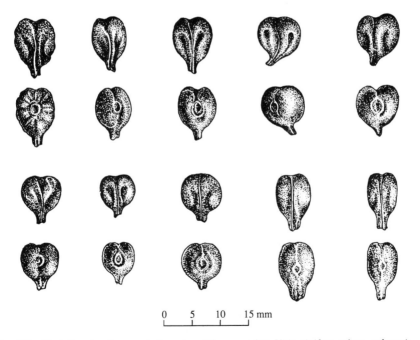

0   5   10   15 mm

Fig. 39b. Variation in pip morphology in wild grape vine *Vitis vinifera* subsp. *sylvestris*. (Zaprjagaeva 1964, plate 286.)

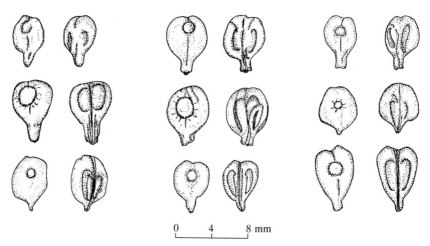

0   4   8 mm

Fig. 40. Carbonized pips of grape vine, *Vitis vinifera*. Early Bronze Age Jericho. (Hopf 1983.)

incomplete and the boundaries between the main morphological forms are blurred, making a clear-cut delimitation of species in *Vitis* a difficult or even an impossible task. Before the colonization of North America, viticulture in Europe and western Asia was restricted to the gene-pool present in the *V. vinifera* complex but, in modern times, several wild species native to America have been used either as additional genetic sources for breeding of new varieties of grape vine or as hardy stocks for grafting.

Archaeological evidence

Berries of wild *Vitis* were collected long before the domestication of the plant. Carbonized pips of grapes and occasionally also berries have been discovered in numerous prehistoric sites in Europe, particularly in northern Greece, Yugoslavia, Italy, Switzerland, Germany, and France (for enumeration of finds, see Renfrew 1973, p. 127; Riviera Núñez and Walker 1989, p.216). In this early material the morphology of the pips conforms with that of local *sylvestris* grapes. The remains antedate viticulture, and represent material collected from wild plants. Several dozens of pips discovered in late Neolithic (4500 bc) Dhali Agridhi, Cyprus (Stewart 1974) also seem to represent wild grape vines.

The earliest definite signs of *Vitis* cultivation come from Chaleolithic and from Early Bronze Age sites in the Levant. Charred pips were recovered from Chaleolithic Tell Shuna North in the Jordan Valley (R. Neef, personal communication). Pips (Fig. 40) and parched berries containing two to three seeds were discovered in Early Bronze Age (*c.* 3200 bc) Jericho (Hopf 1983). These finds have been complemented by remains found in Early

Bronze Age Lachish (Helbaek 1958), Numeira (McCreery 1979), and Arad (Hopf 1978a). Arad also yielded two samples of charred wood; and in Numeira numerous pips and hundreds of whole berries were excavated. The fruits are small and the pips roundish and relatively short-beaked. But since wild *Vitis* is absent today in Jordan and Judea, and is unlikely to have grown wild in these areas in the second half of the 4th and in the 3rd millennia bc, the combination of pips and charred wood provides solid proof that there and then *Vitis* was cultivated. Additional finds come from Kurban Höyük, Urfa district, south Turkey (Miller 1986, 1991). In this site pips of grapes were recorded from 5 per cent of the samples taken from late Chalcolithic beds. Their occurrence increased to 10 per cent of the samples taken from early Early Bronze Age levels. They became very frequent (66 per cent) and come in masses in mid–late Early Bronze Age. This profile seems to reflect a rapid growth of viticulture in the north Levant at the start of the Bronze Age. Finally, infra-red spectroscopy detection of tartaric acid deposits on the inner surface of a large jar recovered from Godin Tepe, western Iran (Badler *et al.* 1990) is very suggestive. It shows that wine was produced in the Near East in the Late Uruk period (middle of the 3rd millennium bc). From the second half of the 3rd millennium bc on, fresh grapes, raisins, and wine are also repeatedly recorded in Mesopotamian cuneiform sources (Postgate 1987).

Evidence for early viticulture and wine-making comes also from ancient Egypt. Remains of grapes and signs of wine production and/or wine import appear already in Old Kingdom tombs (Lauer *et al.* 1950; Germer 1989b). Among the earliest finds are small carbonized pips discovered in 1st dynasty (*c.* 2900–2700 bc) graves in Abydos and Nagada; and a rich find of fragments of raisins found in 3rd dynasty Djoser pyramid at Saqqara (*c.* 2600 bc). Wine jars and wall paintings of grapes appear from Old Kingdom times on. As Egypt lies already far outside the distribution range of wild *Vitis*, the grape had to be introduced to the Nile Valley as a cultivated plant. As argued by Stager (1985), the most plausible source for such introduction is the Levant.

Climatically, Egypt lies already outside the optimal range for *Vitis*; and viticulture does not succeed well in most parts of the hot Nile Valley. It is probable that in dynastic times raisins and wine were largely imported into Egypt; and grapes were planted only as a luxury crop, and were restricted mainly to the cooler delta area.

In the Aegean area definite signs of grape-vine cultivation appear somewhat later. In Thessaly and Macedonia pips become so common at several late Neolithic (4300–2800 bc) sites that Kroll (1991) remarks that their sheer numbers could suggest cultivation. However, the earliest convincing evidence comes from Early Helladic IV (2200–2000 bc) Lerna in south Greece, where numerous pips approaching cultivated *vinifera* in their breadth-to-length ratio were recovered (Hopf 1961b). Additional remains conforming in their shape to the cultivated grape vine were recovered from Middle and from Late Bronze Age sites (Hansen 1988). They include a rich find of several hundreds of elongated *vinifera*-type pips from Late Helladic (1600–1000 bc) Kastanas (Kroll 1983). Parallel to the situation in the Levant the development of grape-vine cultivation in the Helladic and Minoan cultures is indicated also by the presence of presses, and the appearance of specific wine jars and wine cups.

Archaeobotanical documentation of Bronze Age of *V. vinifera* is also available from Transcaucasia (Wasylikowa *et al.* 1991, p. 235). Pips showing the morphology of cultivated forms start to appear in Georgia in Early Bronze Age. They become common in Middle and Late Bronze Age (2nd and 1st millennia bc). Also in Armenia, *vinifera*-type pips were frequently retrieved, from the 2nd millennium bc onwards.

From the end of the 3rd millennium bc on there are convincing signs of grape-vine cultivation in Baluchistan. Wood charcoal of *Vitis* was retrieved from the latest beds (Period VII, *c.* 2000 bc) in Mehrgarh, Pakistan (Thiebault 1989) and pips were found in contemporary, neighbouring Naushario (Costantini and Costantini-Biasini 1986). Similar to Egypt, this part of Baluchistan lies far away from the wild habitats of *V. vinifera*. Again the finds (particularly the charcoal) indicate imported cultivation.

Viticulture was apparently introduced to the west Mediterranean basin and to the Black Sea area by Phoenician and Greek colonists (Stager 1985). The Romans brought this crop to temperate Europe (Loeschke 1933; König 1989).

In summation, the situation in *Vitis* parallels that found in *Olea*. The earliest archaeological indications of viticulture come from areas close to the eastern shore of the Mediterranean Sea. In this general territory we still find wild *sylvestris* forms that could have been used for developing the cultivated grape. But while *oleaster* olives are confined to the typical Mediterranean climate, *sylvestris* vines thrive in somewhat cooler and more mesic conditions. These adaptations are paralleled under cultivation. We are led to conclude that the Levant is the most probable area in which *Vitis vinifera* domestication could have been initiated.

Fig: *Ficus carica*

The fig, *Ficus carica* L., is the third classical fruit tree associated with the beginning of horticulture in the Mediterranean basin (Zohary and Spiegel-Roy 1975). Figs seem to have been part of food production in this area since at least the Early Bronze Age, providing fresh fruit in summer and storable, sugar-rich, dry figs all the year round. The fig is a relatively fast-growing fruit crop. Production starts 3–4 years after planting.

Early fig cultivation was centred in typical Mediterranean environments, in close association with the olive and grape vine. Figs are dioecious and cross-pollinated. In the wild they depend entirely on sexual reproduction from seed. Under cultivation their propagation is vegetative. The grower maintains desired genotypes by rooting of cuttings of winter-dormant twigs and occasionally by grafting. Since *F. carica* is bisexual, female fruit-bearing clones are propagated. Such cultivars are extremely heterozygous and when progeny-tested, they manifest a wide segregation, and most of the seedlings are economically useless. The majority of present-day cultivars, known as 'common figs', are parthenocarpic. They were selected for their ability to produce sweet and large fruits without pollination. A second group of highly appreciated cultivars, the Smyrna-type figs, still require pollination for fruit development (Condit 1947). A single dominant mutation determines the change (under domestication) to parthenocarpy.

The reproductive biology of *F. carica* is complex (Storey 1976; Galil and Neeman 1977) and based on (i) a highly specialized inflorescence (the syconium) which the growers refer to as the fig's 'fruit' (ii) the presence of two sex forms of which the female morph is known as the true fig and the male as caprifig, and (iii) an elaborate symbiosis between the plant and its pollinator, the fig wasp *Blastophaga psenes*.

The syconium, unique to the genus *Ficus*, is a fleshy flowering branch transformed into a hollow receptable which bears numerous minute flowers on its inner surface and is open to the outside by a narrow orifice (ostiole). The true fruits are small druplets or 'seeds' each developing in a female flower inside the syconium.

Young syconia on the female 'true figs' contain only pistillate, female, long-styled flowers. In contrast 'caprifigs' produce spongy, non-palatable syconia containing both staminate male flowers and modified short-styled female flowers. The latter do not set seeds but serve for the multiplication of the fig wasp: they are adapted for oviposition and nourish the *Blastophaga* larvae by turning into galls when eggs are laid in them. True figs produce their main crop of 'fruits' in late summer. Caprifigs usually bear three crops of syconia during the year: over wintering *mamme*, numerous *profichi* which develop during spring, and *mammoni* which ripen in autumn. In all three of them, the *Blastophaga* wasps develop synchronously with the syconia. Many of the pollen-carrying female wasps emerging from mature syconia of caprifigs, particularly those from the abundant *profichi* produced in late spring, do not land only on new young caprifig syconia, but are attracted also to the numerous young female syconia borne at this time by true figs (Galil and Neeman 1977). They enter the syconia through the orifice, become trapped in them, bring about pollination and perish. Since the wasps are unable to insert their eggs into the long-styled female flowers, the maturing 'fruits' of the true fig do not harbour *Blastophaga* larvae even after having been visited by the insects.

The cultivator growing Smyrna-type figs makes use of the symbiosis between the fig and the fig-wasp by artificial pollination known as caprification: twigs with mature *profichi* are collected from wild caprifigs in early summer and suspended on the true fig trees in the plantation. The pollinating wasp is thus brought near to the female syconia. Caprification is an ancient procedure. It was practised in Greek and Roman times (White 1970, p. 228) and probably even earlier.

Carbonized small seed constitute the bulk of fig tree remains retrieved in archaeological digs. Some sites have also yielded charred, whole, dry figs. The archaeological record of figs is still sparse. It is also likely to be biased since the pips of *F. carica* are small and may be overlooked by the excavator.

Wild ancestry

The cultivated fig tree shows a close morphological resemblance, striking similarities in climatic requirements, and tight genetic interconnections with an aggregate of wild and weedy forms which are widely distributed over the Mediterranean basin (Map 17). Botanists (see for example, M. Zohary 1973, p.631) regard these spontaneous figs as the wild progenitor of the cultivated fruit tree and place them within *F. carica* L. Wild figs (Fig. 41) grow mainly in the low altitudes of the Mediterranean *maquis* and *garrigue* formations. They occupy rock crevices, gorges, stream sides, and similar primary habitats. They are often complemented by a wide range of feral types occupying secondary, man-made habitats such as edges of plantations, terrace walls in cultivation, ruins, collapsed cisterns, cave entrances, etc. Frequently these 'weedy' types seem to be derived from seed produced by local cultivated clones which were pollinated by the adjacent wild-growing caprifigs (Zohary and Spiegel-Roy 1975). Wild and weedy populations of *F. carica* are normally composed of an equal number of female true fig and male caprifig trees. Wild females usually bear relatively small, hardly edible fruits.

The Mediterranean wild-feral-cultivated *F. carica* complex is closely related to a group of non-Mediterranean wild, deciduous *Ficus* types distributed south and east of the Mediterranean region (Warburg 1905; M. Zohary 1973, p. 630; Browicz 1986). Taxonomically, all of them form a single natural group (series *Carica* in section *Eusyce*) within the genus *Ficus*. The latter is an enormous genus comprising some 1000 species distributed mostly in the tropics. Tall, large figs grow in the lower zone of the mesic, deciduous forests of the Colchic (Black Sea) district of northern Turkey and the Hyrcanic (south Caspian Sea) district of Iran and adjacent Caucasia. These forest types intergrade with the typical Mediterranean *F. carica*. Most authors include these mesic, wild forms within *F. carica* L. However, several Russian botanists (see, for example, Zhukovsky 1964), treat them as two independent species: *F. colchica* Grossh., and *F. hyrcanica* Grossh.

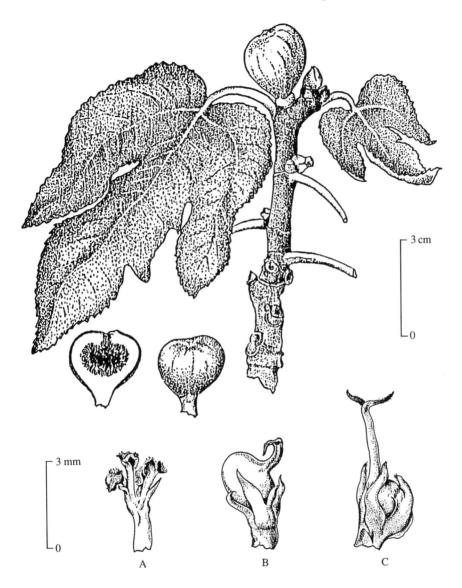

Fig. 41. The fig *Ficus carica*. Upper part: a twig with a flowering syconium. A – Staminate flower with 4 anthers. B – Modified short styled female flower. C – Long styled female flower. (With kind permission of J. Galil.)

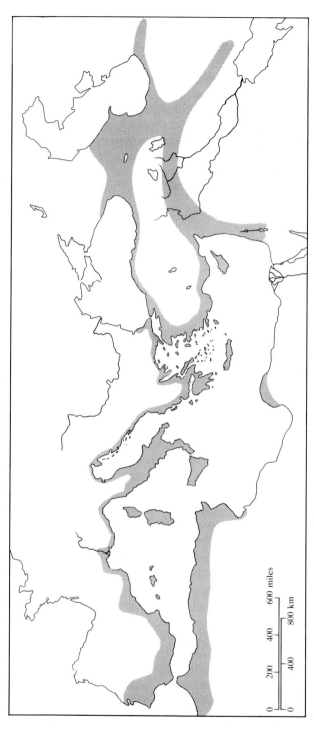

Map 17. Distribution of wild fig, *Ficus carica*. (The distribution in the eastern half of the map is based mainly on data from Browicz 1986.)

600 miles

800 km

200   400   600

0   400   800

Other members of the series *Carica* are warm climate, xeric shrubby types distributed outside the traditional area of fig cultivation: *F. johannis* Boiss. (syn. *F. geraniifolia* Miq.) in the Zagros Mountains and southern Iran; *F. virgata* Roxb. in Afghanistan; *F. pseudosycomorus* Decne in the Negev, Sinai, and Egypt, and *F. palmata* Forssk. in Yemen, Somalia and Ethiopia. Some of these wild figs are interconnected by intermediate forms and should perhaps be given only a status of eco-geographic subspecies. But none of them, with the exception perhaps of the tall Colchic forest type, have established noteworthy contacts with the crop complex.

Several of these non-Mediterranean wild figs were genetically tested (Storey and Condit 1969) and found to have a chromosome complement identical to that of the cultivated fig ($2n = 26$). All members of series *Carica* seem to be fully interfertile with the crop complex but adapted to different ecological regimes. The close affinities between the members of the *Carica* series are also indicated by the behaviour of *Blastophaga psenes*. At least in experimental plots the insect moves freely between the Mediterranean and non-Mediterranean types.

Archaeological evidence

Remains of fig pips have been retrieved from several east Mediterranean Neolithic sites such as Early Neolithic (7800–6600 bc) Tell Aswad, Syria (van Zeist and Bakker Heeres 1979), PPNA (*c.* 7000 bc) Jericho (Hopf 1983), ceramic Neolithic Dhali Agridhi, Cyprus (Stewart 1974), and Neolithic Sesklo in Greece (Kroll 1981a). Fig pips are also recorded from Bronze Age Vallegio, north Italy (Villaret-von Rochow 1958). It is impossible to distinguish between pips of wild and cultivated figs; and the excavated material can be interpreted as either. However, we are inclined to regard all these finds as representing collections from the wild. Very probably the fig was brought into cultivation at the same time as its two horticultural companions: olive and grape vine.

As to later sites, carbonized pips of figs were uncovered in Chalcolithic Tell Shuna North, Tell Abu Hamid and Tuleilat Ghassul (Neef 1990), in Chalcolithic and Bronze Age Jericho (Hopf 1983) and in Early Bronze Age Bab edh-Dhra (McCreery 1979) in the Dead Sea basin. Very small pips as well as several whole fruits were retrieved from late Neolithic and younger horizons of Lerna (Hopf 1961b), Greece. Cuneiform sources indicate that figs were grown in Mesopotamia from the second half of the 3rd millennium bc on (Postgate 1987). In Egypt, the earliest record on figs is from the 3rd dynasty, about 2750 bc, and a beauti-ful drawing of a fig harvest was found in Khnumhotep's grave (about 1900 bc) in Beni Hasan (Darby *et al.* 1977). Contemporary indica-tions of fig cultivation are available also from Syria (Stager 1985). They show that since the Bronze Age, figs accompanied olives and grapes as main horticultural elements of the rain-dependent agriculture in the Mediterranean basin.

When the data available from archaeology and the information extracted from living plants are combined, we are led to conclude that, in this species too, the earliest cultivation was practised in the eastern part of the Mediterranean basin. This very territory harbours the closest wild relatives of the cultivated fig. Thus the wild *F. carica* forms of the Levant, southern Turkey and the Aegean belt can be regarded as representatives of the ancestral wild stock from which early fig domesticants were derived. The main changes in this fruit tree under domestication were the shift to vegetative propagation of female clones, the increase of the size and sugar content of the syconium, the introduction of artificial pollination or caprification, and the selection for parthenocarpic clones in which pollination is not necessary. Such clones were already part of horticulture in classical times.

Sycamore fig: *Ficus sycomorus*

The cultivation of the sycamore fig, *Ficus sycomorus* L. has been almost exclusively an Egyptian speciality. Compared with the true fig, *F. carica*, it is a much taller tree but it produces smaller and inferior fruits. Since early dynastic times the sycamore fig was (and still is) a valued domesticant in the lower Nile Valley (Täckholm 1976). It yields a steady supply of sweet syconia, and highly appreciated timber. Outside Egypt the sycamore fig was much less esteemed, though it was cultivated in the warmer parts or Israel and to a lesser extent also in several other locations on the shores of the Mediterranean Sea (e.g. Lebanon, Cyprus, Tunisia). As in other members of *Ficus*, the reproduction of the wild plant is from seed and seed setting depends on pollination by a specific symbiotic wasp, *Ceratosolen arabicus* Mayr. In contrast, cultivation is based on clonal propagation (rooting of twigs), and fruit production no longer depends on the pollinator: some clones are parthenocarpic and set fruit without fertilization. In others, fruit maturation is artificially induced by gashing the surface of the young syconia. This is an old tradition in Egypt and in Israel (Galil *et al.* 1976).

The wild *F. sycomorus* is widely distributed in eastern Africa from the Sudan to South Africa, with an extension to Yemen. The tall trees, growing near streams and in beds of ephemeral watercourses, constitute a conspicuous component of savanna landscapes. Throughout this area, *F. sycomorus* reproduces sexually and pollination is effected by the *Ceratosolen* wasp. As far as we know (Galil *et al.* 1976), spontaneously reproducing sycamores, as well as their pollinating wasp, are today absent in Egypt.

Remains of *F. sycomorus* appear in quantity in Egyptian excavations from the start of the 3rd millennium bc on (Galil *et al.* 1976; Germer 1985, p. 26) The fruit and the timber, and sometimes even the twigs, are richly represented in the tombs of the Early, Middle, and Late Kingdoms. In numerous cases the parched sycons bear characteristic gashing marks indicating that this art, which induces ripening, was practised in Egypt already in ancient times.

There is no doubt that Egypt was the principal area of sycamore fig development. In spite of the absence of spontaneous plants in Egypt now, it seems that *F. sycomorus* was brought into cultivation in Egypt, and that the wild stock and the wasp have become extinct in this country since then (Galil *et al.* 1976). Wild sycamores and their symbiotic wasps thrive today in nearby Sudan.

Date palm: *Phoenix dactylifera*

The date palm, *Phoenix dactylifera* L., ranks among the first fruit trees which were taken into cultivation in the Old World. It was already part of the Near East food production in the Chalcolithic period (Zohary and Spiegel-Roy 1975). Compared with the three other classic fruit trees (olive, grape vine, fig), the date palm requires a much warmer and drier climate. High temperatures and low air humidity are particularly important for fruit setting and fruit ripening. Date cultivation is therefore centred in the deserts south of the Mediterranean Sea and in the southern fringe of the Near East, from south Iran in the east to the Atlantic coast of North Africa in the west, between 35°N and 15°N latitude. This is an almost rainless belt, and cultivation depends on a steady water supply, either by means of irrigation or the presence of a high level of ground water. Date palms can withstand considerable salinity and they also thrive when irrigated by brackish water. They are very productive and the fruit yield may be as high as 100–200 kg per tree. All over these warm desert territories, the sugar-rich fruits (sugar content 60–70 per cent) serve as an important food. The trunks are cut and split longitudinally into halves or quarters and serve as beams for building, the leaves are extensively used for roofing, matting, and basketry, and the fibres of the bark for ropes. Dates start to bear fruits 4–5 years after planting and reach full production at 8–10 years. For details on date palm cultivation and on the flower and fruit biology of this crop consult Nixon (1951), Dowson (1982), and Reuveni (1986).

Phoenix dactylifera is characterized by its tall trunk (up to 25 m), its large, leathery, feather-shaped leaves and its ability (when young) to produce basal suckers. The vegetative propagation of clones practised under domestication depends on these suckers. The plant is dioecious. Wild populations, as well as seedlings derived from cultivated clones, consist of an equal proportion of female and male individuals. This is brought about by a single gene, or block of genes, determining sexuality.

The male morph is heterozygous and the female homozygous recessive. Pollination is carried out mainly by the wind. Under domestication, the number of male individuals required for ensuring pollination and fruit setting in the plantation has been reduced to one male for 25–50 females by introducing artificial, hand-carried pollination. This tradition is very old (Schiel 1913; Pruessner 1920) and it is probable that it was already practised in Mesopotamia in Hammurabi's time.

Wild ancestry

The wild stock from which the cultivated date palm could have been derived is already satisfactorily identified (Zohary and Spiegel-Roy 1975). The crop is closely related to a variable aggregate of wild and feral palms distributed over the southern, warm, and dry Near East as well as the north-eastern Saharan and north Arabian deserts. These spontaneous dates show close morphological similarities and parallel climatic requirements with the cultivated clones. In addition, they are fully interfertile with the cultivars and are interconnected with them by occasional hybridization. Botanists place these wild dates with *P. dactylifera* L. The wild forms produce basal suckers just as the cultivated varieties, but differ from the domesticated clones by their smaller fruits. These contain relatively little pulp and are frequently non-palatable or even indigestible. Thus, in the date palm also, domestication has led to the increase in fruit size and pulp quality. Wild and cultivated date palms also differ in their mode of reproduction. Sexual reproduction is the rule in wild stands while cultivation has brought a shift to vegetative propagation, i.e. to the 'fixation' of desired highly heterozygous female clones. *P. dactylifera* was well suited for this shift since it produces easily transferable suckers.

Spontaneously growing dates occur in almost the entire area of date cultivation (Map 18) and they frequently represent secondary escapees. However, in some areas in the Near East and very probably also in the north-eastern Sahara, in Arabia, and in Baluchistan, dates are genuinely wild and occupy primary niches. Prominent among these locations are lowland Khuzistan, the southern base of the Zagros Range facing the Persian Gulf, and the southern part of the Dead Sea basin. In these warm, dry territories, wild *dactylifera* palms with their characteristic small and mostly inedible fruits thrive in gorges, wet rocky escarpments, seepage areas in wadi beds and near brackish springs where they constitute a conspicuous element in the vegetation of these areas. In *P. dactylifera* we are therefore faced with a variable complex of wild forms, segregating escapees, and cultivated clones which are all genetically interconnected by occasional hybridization. It is often difficult to decide whether non-cultivated material is genuinely wild or whether it represents weedy forms or secondary seedlings derived from cultivated clones. But though it is impossible to

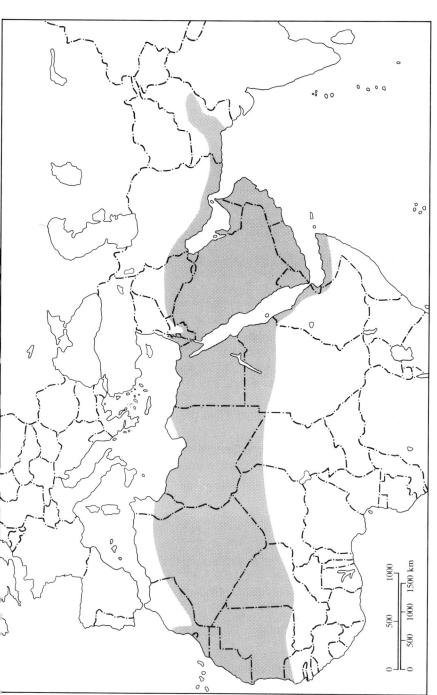

Map 18. Distribution of wild and feral forms of the date palm, *Phoenix dactylifera*. In Morocco they sometimes intergrade with the local *P. atlantica*; in south Arabia, Sudan, Ethiopia, and Senegal with native *P. reclinata*; and in the Indus basin with *P. sylvestris*.

delimit the pre-agriculture distribution of the wild date, there can be little doubt that the wild *dactylifera* forms are indigenous to the warm and dry parts of the Near East.

Phoenix dactylifera is the main Near Eastern wild representative of its genus which comprises some 12 species distributed over Africa and south Asia (Corner 1966; Munier 1973). (The only other wild date which occurs in the east Mediterranean basin is *P. theophrastii* Greuter, a narrow endemic confined to Crete and several spots in coastal south-west Turkey.) Several species of Phoenix were tested cytogenetically. All show a constant chromosome complement ($2n = 36$) and are fully interfertile with one another. As in the case of *Vitis*, they are more or less reproductively isolated from one another by means of geographical and ecological barriers. Three wild *Phoenix* species grow on the fringe of traditional date cultivation in the Old World and, in these areas, they may have enriched the gene-pool of the cultivated fruit trees through spontaneous hybridization: *P. atlantica* A. Chev. grows near the Atlantic shore of North Africa, and it is apparently involved in the formation of some Moroccan date cultivars; bushy *P. reclinata* Jack., occurs in south Arabia and in Africa south of the Sahara, and it probably hybridizes with *P. dactylifera* on the southern fringe of date cultivation, finally on its eastern border, the Indus Valley, the cultivated date comes in contact with the wild *P. sylvestris* Roxb., a non-suckering, tall, 'rain palm' not adapted to deserts but to wetter, tropical climates. *P. dactylifera* occasionally hybridizes also with this palm.

Archaeological evidence

Archaeological sites with date palm remains are so far few, but this may reflect the fact that most excavations in the Near East and the Mediterranean basin were conducted in areas too cold for date cultivation.

The very few stones of date palm reported from Egypt, Iran, and Pakistan and dated to the 6th and 5th millennia bc, probably represent material collected from the wild. The earliest remains of what seem to be cultivated dates have been found by Seton Lloyd in the Ubaidian horizon (*c.* 4000 bc) at Eridu, Lower Mesopotamia ("buckets of stones", R. J. Braidwood, personal communication). In Jordan, carbonized stones (Fig. 42) were discovered in the classic Chalcolithic (3700–3500 bc) site of Tuleilat Ghassul (Zohary and Spiegel-Roy 1975). Date kernels were also retrieved from the Chalcolithic bed (3500–3200 bc) of the 'Cave of the Treasure' in Nahal Mishmar (Bar-Adon 1980) and a single date stone is available from Early Bronze Age Jericho (Hopf 1983). Date stones were also found among offerings deposited in a tomb in the 'Royal Cemetry' at Ur, lower Mesopotamia and dated to the late 3rd millenium bc (Ellison *et al.* 1978). Imprints were also found in *c.* 2500 bc Hili Oasis, Oman peninsula (Cleuziou and Costantini 1982).

Fig. 42. Carbonized stones of dates. *Phoenix dactylifera.* Chalcolithic Tuleilat Ghassul, Jordan. (Zohary and Spiegel-Roy 1975.)

From the Bronze Age onwards, date cultivation seems to have been well established in the warmer sections of the Near East. Dates are frequently mentioned in Sumerian and Akkadian cuneiform sources – from the second half of the 3rd millennium bc onwards (Postgate 1987). Postgate (1980) stresses the fact that although archaeological evidence is yet insufficient, the written documentation shows that at the latest by the Late Uruk period (*c.* 3250 bc) the cities of Sumer possessed flourishing date plantations.

Remains of dates appear in Egypt in the Middle Kingdom (Täckholm and Drar 1950); and Täckholm (1976) stresses that they become very common in the New Kingdom. In this period, remains of date fruits occur 'in almost every second tomb; in all forms, all sizes, and all colours'. In the 3rd millennium bc date cultivation is also practised in Baluchistan (Costantini and Costantini-Biasini 1985). In classical times, dates appear to be an important food element in lower Mesopotamia, the lower Jordan Valley, the Nile Valley, and the desert oases of North Africa, Arabia and Baluchistan.

The available archaeological data, as well as the information on the living date palm, seem to focus on the Near East as the initial place of

P. dactylifera domestication. The same area which furnishes the earliest indications on date palm cultivation also harbours wild forms from which the first cultivated clones could have been obtained. Very probably the date palm was first brought into cultivation somewhere in the lower Mesopotamian basin, or in some oases in the southern fringe of the Near Eastern arc.

Pomegranate: *Punica granatum*

The pomegranate, *Punica granatum* L., is an appreciated minor crop in traditional Mediterranean horticulture. It is a deciduous bush or small tree with conspicuous red flowers and large (6–12 cm across) fruits characterized by a leathery rind, persistent crown-like calyx, and numerous seeds covered with a juicy flesh which can be eaten fresh, or whose juice can be extracted, and can also be fermented into wine.

The wild ancestor of the cultivated pomegranate is satisfactorily identified (Zohary and Spiegel-Roy 1975). Wild forms of *P. granatum* grow in masses in the south Caspian belt, in north-eastern Turkey, and in Albania and Montenegro. Domestication brought about an increase in fruit size, and a shift from sexual reproduction to clonal propagation. Desired genotypes are easily maintained by rooting of winter-dormant suckers. Cultivars set fruit by self-pollination.

The pomegranate seems to belong to the early ensemble of cultivated fruit trees in the Old World. As well as supplying fresh fruit they were also considered to be a symbol of fertility. Carbonized pips and fragments of pomegranate peel have been obtained from Early Bronze Age Jericho (Hopf 1983) and Arad (Hopf 1978a). They also appear in Late Bronze Age Hala Sultan Tekke, Cyprus (Hjelmqvist, 1979b) and in Late Bronze Age Tiryns, Argolis (Kroll 1982). Pomegranates are recorded in several ancient Mesopotamian cuneiform sources – from the second half of the 3rd millennium bc on (Postgate 1987). Schweinfurth (1891) identified pomegranate among the vegetable remains of the 12th dynasty (1970–1800 bc) Drah Abu el Naga, Egypt. A large dry pomegranate was found in the tomb of Djehuty, the butler of Queen Hatshepsut, dating from about 1470 BC (Hepper 1990, p. 64). Pomegranates do not occur in a wild state in the south Levant so the occurrence of *Punica* remains in Early Bronze Age in Israel strongly suggest well established cultivation.

Apple: *Malus pumila*

The apple, *Malus pumila* Mill. (Syn. *Pyrus malus* L., *Malus domestica* Borkh.), is the most important fruit crop of the colder and temperate parts of the Old World (Brown 1975; Watkins 1981). Apples are extensively grown in Europe, Turkey, north Iran, Caucasia, and central Asia. They thrive in regions where winters are sufficiently cool to provide the trees with a chilling phase necessary to break bud-dormancy. The fruit is a pome, i.e. a 'false fruit', the greater part of which is formed by the receptacle of the flower and not by the ovary. In the core there are five leathery chambers or loculi, each normally containing two seeds.

Apples under cultivation are very variable. Some 2000 distinct cultivars are known and they vary in fruit size (3–12 cm in diameter), colour (red, yellow, green), shape, and in the texture and taste of the pulp (sweet to acid). The majority of present-day cultivars belong to the group of 'dessert apples' which bear relatively sweet fruits that are consumed fresh. 'Cooking apple' varieties produce harder and often larger fruits. Other varieties are 'cider apples', selected for the extraction of juice for fermentation of vintage quality cider. Less 'fancy' types of this alcoholic beverage can be prepared by fermenting ordinary dessert apples or wild 'crab apples'. Some apples are grown for their ornamental features and certain other clones serve as root-stocks on which the fruit cultivars are grafted.

Most apples are fully or at least partly self-incompatible and in order to set fruit, have to be pollinated by a different genotype (clone). The majority of the cultivars have a diploid ($2n = 34$) chromosome constitution, but numerous clones are triploid ($2n = 51$) and consequently develop only a few seeds in their fruit. Only a few varieties are tetraploid ($2n = 68$).

Cultivated apples segregate widely when grown from seed, and most progeny are economically worthless. As in many other fruit trees, cultivation depends on vegetative propagation of clones. Since the majority of apples cannot be reproduced from cuttings or suckers, their cultivation depends totally on grafting. Propagation from suckers is possible only in a few small-fruited cultivars of western Asia, e.g. the 'Hashabi' group in the Near East. Carbonized fruits, occasional pips, and charred wood occur in archaeological sites.

Wild ancestry

Malus Mill. is a genus of 25–30 species distributed over the temperate belt of Europe, Asia, and North America. Cultivated apples are closely related and fully interfertile with a variable group of wild and feral apples which are widely distributed over the temperate and cool forest areas of Europe and Asia. Wild apples reproduce entirely from seed, are self-incompatible, and are characterized by small fruits (1.5–3.0 cm in diameter), which are very variable in their shape, colour, and taste. They show a marked differentiation into eco-geographical races which have been ranked by taxonomists as independent species. The widespread European wild 'crab apples' are usually referred to as *M. sylvestris* (L.) Mill. Forms growing in Anatolia, north Iran, and the Caucasus are either regarded (Browicz 1972) as a distinct regional subspecies, *M. sylvestris* subsp. *orientalis* (Uglitzkich)

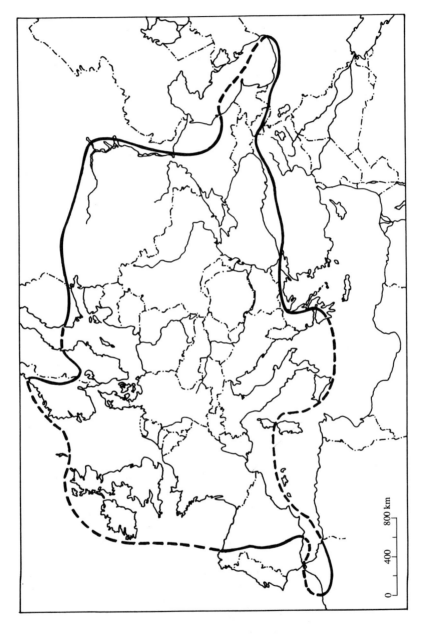

Map 19. Distribution of wild apple, *Malus sylvestris* (including subsp. *orientalis*). Based on information kindly provided by K. Browicz.

Browicz, or are known as *M. orientalis* Uglitzkich. The central Asiatic wild apples are either grouped in *M. sieversii* (Ledeb.) M. Roem. or split even further by Russian botanists who recognize several additional local species: *M. kirghizorum* A. & Fed. and *M. turkmenorum* Juz. & M. Pop. The eastern Siberian and north Chinese forms are placed into *M. prunifolia* (Willd.) Borch. Variation in apples is further complicated by the fact that in numerous places in Europe, the Near East, Caucasia, and central Asia, apple cultivation has been practised in areas supporting wild *Malus* populations. Furthermore, in some regions, e.g. in Anatolia and the Balkans, wild apples were commonly used where they grew, by grafting them *in situ* with scions from the cultivars. Under such conditions, spontaneous hybridization between tame and wild apples occurs frequently, resulting in the formation of extensive swarms of secondary hybrid derivatives. In recent times, variation patterns in apples have been further complicated by crossing European apples with numerous wild and cultivated forms from other geographic origins (including some wild North American *Malus* species). This has produced a whole array of new hybrid combinations.

Archaeological evidence

Apples were, without doubt, collected from the wild long before their domestication and before the establishment of vegetatively propagated clones. Numerous carbonized remains of small apples (20–27 mm in diameter), often cut into halves for parching were found in Neolithic and Bronze Age Switzerland (Schweingruber 1979). These remains correspond very well to the wild apple forms which grow today in that area. Similar remains of small apples were retrieved in numerous Neolithic and Bronze Age sites in Europe, for example in Yugoslavia, Hungary, Czechoslovakia, Austria, Germany, and Denmark (for an enumeration of finds, see Hopf 1973c); they too have been collected from the wild.

Small apples, cut transversely in half, threaded and dried on strings, were found among offerings deposited in a tomb in the 'Royal Cemetry' at Ur, lower Mesopotamia, and dated to the Sargonid period, late 3rd millennium bc (Ellison *et al.* 1978). This discovery confirmed descriptions of strings of apples in cuneiform texts from the mid 3rd millenium bc. Since apples do not grow wild in warm and dry Mesopotamia, such small apples could represent a long distant import of wild apples or early attempts of apple cultivation. Because *Malus* does not lend itself to simple vegetative progatation, apple cultivation in these early times would have been necessarily based on seed planting. In other words, on handling of variable segregating individuals resembling 'crab apples'. This is indeed what the small, stringed apples look like. Yet even such low-grade apples seem to have been appreciated in ancient Babylonia, and were used for offering.

A similar early indication of apple cultivation comes from Kadesh-Barne'a, a Judean outpost in the tenth century BC on the border between the Negev and Sinai. Here several dozens of well-preserved carbonized fruits have been discovered (Rudolph Cohen, personal communication). Wild apples do not occur in this area and their southernmost limit today is Turkey. These remains, therefore, strongly suggest that apples were cultivated in the oasis of Kadesh Barne'a.

Whatever the nature of the Ur and Kadesh Barne'a finds, one fact is clear: the apple did not evolve into a major fruit crop in Bronze Age times. While the cultivation of the olive, the grape vine, and the date palm accelerated dramatically in the 3rd and 2nd millennia bc, there are no signs of parallel development in the apple. This tree seems to have obtained its significant role in Old World horticulture only in classical times. This development is apparently linked to the introduction of grafting. The Greeks were already familiar with the art of grafting (White 1970, p.248) and Theophrastus writes about the maintenance of apple clones by this sophisticated method of vegetative propagation.

In conclusion, we still know very little about the time and place of apple domestication except that in classical times, apples were already extensively grown in the Old World. Wild apples are widely distributed over Europe and west Asia where they have diverged into numerous types and closely related species which still retain full interfertility. Apple cultivation is superimposed on this wild background. It is therefore futile to try to delimit the area of initial domestication on the basis of the evidence from the living plants. Apples could have been brought into cultivation anywhere in the temperate areas of Europe and western and central Asia. Very probably, exceptional *Malus* individuals were picked up not once and in a single place, but many times and in several areas. Furthermore, many cultivars are hybridization products combining genes from several distinct geographic sources. Nevertheless, most of the old type varieties in Europe and the Near East have their closest morphological affinities with the wild *M. sylvestris* and this undoubtedly is the principal, but not the only wild source, from which the old cultigens in these regions have evolved.

The archaeological finds make it clear that although apples have been extensively collected from the wild during Neolithic and Bronze Age times, the first signs of substantial apple cultivation appear much later than those of olive, date palm, or grape vine horticulture. Grafting is apparently a key element in this delay. Because apple cultivation depends almost entirely on this method, it could only have been practised after this technology became known in the Near East and Europe. Where and when grafting was invented is not yet clear, although its initiation was very probably not in the area of Mediterranean horticulture (see pp.135–6).

Pear: *Pyrus communis*

The cultivated pear, *Pyrus communis* L. (syn. *P. domestica* Med.), is second to the apple in its contribution to fruit production in the cool and temperate parts of Old World agriculture (Layne and Quamme 1975; Watkins 1981). Frequently, pears are close companions of apples, and both fruit crops thrive in similar environments. Pears, too, require sufficient winter chilling to ensure normal flowering and fruit setting. As in apples, the fruits are pomes in which the greater part is formed by the receptacle of the flower and only the inner 'core' containing the seed develops from the ovary. Usually the pomes have the distinctive pyriform shape of *Pyrus* and their pulp contains stone cells that give pears their characteristic gritty texture.

Cultivated pears are very variable and some thousand distinct cultivars have been named. Like most other rosaceous fruit trees, pears are self-incompatible. Most cultivated clones are diploid ($2n = 34$); some are triploid ($2n = 51$). Some pear cultivars are parthenocarpic, i.e. can set fruit without fertilization and seed development. Pears segregate widely when grown from seed and most of the progeny is worthless. Like apples, the maintenance of desired clones depends entirely on grafting.

Wild ancestry

The genus *Pyrus* L. comprises some 30 species distributed over Europe and Asia (Rehder 1967; Browicz 1993). The cultivated pear is closely related to a variable aggregate of wild pears distributed over the cooler areas of Europe and Asia. The main wild types in this aggregate were formerly treated by botanists as distinct species, though they frequently intergrade and at least some of them conform to eco-geographic races. Most of these wild pears are rather spinescent and bear very gritty small fruits (1.5–3.0 cm in diameter). All are diploid ($2n = 34$), self-incompatible trees; all are interfertile with the cultivated crop. Those which closely resemble the cultivated pears are a group of wild pears distributed over temperate Europe, Caucasia and northern Turkey and known by the names of *P. pyraster* Burgstd. and *P. caucasica* Fed. Spontaneous crosses between these wild pears and cultivated clones are quite frequent. In numerous localities in Europe and western Asia, wild and tame pears are interconnected by feral forms and hybridization derivatives that thrive best at edges of cultivation and in clearings of forests adjacent to cultivation. Because of their tight morphological and genetic affinities with the crop, *pyraster* and *caucasica* pears are now regarded as the main wild stocks from which the cultivated crop has probably evolved. Consequently recent taxonomic treatments of *Pyrus* (Browicz 1972) place them, as wild subspecies, in *P. communis*. (For details of the distribution of wild forms of *P. communis* see Meusel *et al.* 1965, Vol. 2, p. 208; Browicz 1992, p.23 and map 37.)

Several other wild pears that grow in the general area of traditional pear cultivation are interfertile with the cultivated crop. Some or even all of these species have enriched the genetic variation of cultivated pears through hybridization and introgression (Rubzov 1944; Watkins 1981). Prominent among these wild pears are *P. spinosa* Forssk. (= *P. amygdaliformis* Vill.) native to west Turkey, the Aegean basin and the south Balkans; *P. eleagnifolia* Pallas distributed over Turkey and east Bulgaria; *P. salicifolia* Pallas in Caucasia and adjacent areas in Turkey; the more arid *P. syriaca* Boiss. of the Near East 'arc'; and its closely related *P. korshinskyi* Litw. in central Asia. (For details on the distribution of these species see Browicz 1982, pp. 49–52 and maps 79–83.) In earlier times, wild pears were frequently grafted *in situ*. This practice is still common in Anatolia, where farmers frequently spare wild individuals of wild *P. communis*, *P. syriaca*, *P. spinosa* and *P. elaeagnifolia* at the edges of cultivation and inside grain crop fields. These trees are grafted with cultivated clones. Often only one or a few main branches bear scions; the others are left as they are. In such situations, cross-pollination between the two types is almost inevitable.

Finally, it should be pointed out that several wild pears native to northern China and eastern Siberia were apparently brought into cultivation independently and gave rise to the Far East pear cultivars (Layne and Quamme 1975). These include the Chinese sand pear, *P. pyrifolia* (Burm.) Nakai, the Ussurian pear, *P. ussuriensis* Maxim. and the Chinese white pear, *P. bretschneideri* Rehd. Recently, crosses between *P. communis* and these oriental pears resulted in the formation and selection of new and important commercial pear varieties and have caused further extensive genetic fusion in the genus *Pyrus*.

Archaeological evidence

Pears were collected from the wild long before their introduction into cultivation. Carbonized remains of small fruits, sometimes halved and very probably dried, were found in several Neolithic and Bronze Age sites in Europe, e.g. north Italy, Switzerland, Yugoslavia, and Germany (for enumeration of sites, see Hopf 1978b). Similar finds are available from late Neolithic Dimini and Bronze Age Kastanas, Greece (Kroll 1983) and from some Tripolye culture settlements in Moldavia and the Ukraine (Janushevich 1975).

The archaeological evidence up till now does not provide definite clues about the beginning of pear domestication. Reliable information on pear cultivation first appears in the works of the Greek and the Roman writers (Hedrick *et al.* 1921; White 1970). Theophrastus (*c.* 300 BC) describes three cultivated varieties in Greece and also mentions that they were propagated by grafting since propagation by seed results in inferior progeny. Cato (235–150 BC) describes methods for pear cultivation and reported on six pear cultivars grown in his time. Somewhat later, Pliny (AD 23–79) recognized 35 cultivars. The range of fruit characters described by these authors already covers many types of today's pears.

In conclusion, the patterns of variation in Old World pears closely resemble those of apples. Wild pears are widely distributed over Europe and western Asia and manifest considerable divergence into local eco-geographic races and species, though they retain their interfertility. Pear cultivation is superimposed on wild populations and hybridization between wild and tame pears has repeatedly produced secondary hybrid derivatives that tend to colonize disturbed sites. Faced with a crop complex with such a vast distribution and extensive genetic fusion, one can scarcely determine an area of origin of pear cultivation on the basis of the living plants. Various regions in west Asia and in Europe, within the distribution range of *pyraster* and *caucasica* pears, are equally attractive possibilities.

Plum: *Prunus domestica*

Plums rank second to apples and pears in terms of their role in fruit production in the cooler and temperate parts of the world. They can be eaten fresh, cooked, or dried as prunes. The traditional plum of the Old World has been *Prunus domestica* L. (Weinberger 1975; Watkins 1981). This is a variable group of hexaploid ($2n = 48$) cultivars comprising the true or 'European' plums (*P. domestica* subsp. *domestica*), and the damsons, bullaces and the greengages (subsp. *insititia*). Other plums, belonging to the diploid ($2n = 16$) *P. salicina* Lindl. have been taken into cultivation in east Asia and are known as Japanese plums. Until about 200 years ago the European plums and the Japanese plums were separate crops, but in recent years they were repeatedly hybridized with one another, as well as with several other *Prunus* sources such as the American plum *P. americana* Marsh. Many of the modern plum cultivars are hybrid products of this recent genetic fusion (Weinberger 1975).

Plum cultivation depends entirely on grafting. Simpler methods of vegetative propagation, such as rooting of cuttings, do not succeed in this fruit tree. Reproduction from seed is also impractical since cultivars are highly heterozygous and their progeny segregate widely. Most of the seedlings are economically worthless. The majority of hexaploid *P. domestica* clones are self-compatible or at least partly self-compatible. In contrast the diploid Japanese and American plums are self-incompatible.

Spontaneous individuals of hexaploid *P. domestica*, with small (2–4 cm) subglobose fruits, are common in many parts of temperate Europe and Turkey. They thrive in woods, cleared hillsides, edges of cultivation, and hedges. European botanists (see, for example, Webb 1968) regard them as feral or naturalized elements. However, pre-neolithic remains of carbonized plum-stones discovered in the Upper Rhine and the Danube regions closely resemble the stones of present day spontaneous *domestica* plums. These finds suggest that *P. domestica* could have anteceded agriculture and should be regarded as an indigenous element in middle Europe (Werneck and Bertsch 1959).

The 6× European true plums (*P. domestica* subsp. *domestica*) and the 6× damsons, bullaces, and greengages (*P. domestica* subsp. *insititia*) are closely related to a variable aggregate of wild and cultivated plums grouped in *P. divaricata* Ledeb. (= *P. cerasifera* Ehrh.). The wild forms of this aggregate are distributed over the Balkan, Caucasia, and south-west Asia. They are apparently all self-incompatible, reproduce from seed and bear roundish, small (*c.* 2 cm in diameter) green, purple or dark violet fruits which taste very much like the cultivated *domestica* plums. *Divaricata* wild forms fall into several eco-geographic races (Browicz 1972 and personal communication): *P. divaricata* subsp. *divaricata* with glabrous twigs and leaves grows in the more temperate northern territories. *P. divaricata* subsp. *ursina* (= *P. ursina* Kotschy) with a relatively villose lower leaf surface and hairy twigs is native to the Levant and south-east Turkey. Still another pubescent race, *P. divaricata* subsp. *caspica* (= *P. caspica* Kov. & Ekin.) occurs in the Caspian coast of Caucasia and Iran. The cultivated forms of *P. divaricata*, known as myrobalans or cherry plums, are used either as ornamentals or as root-stocks for grafting *domestica* cultivars; and in Caucasia also as fruit bearing 'Alyča' plums. Chromosomally, *P. divaricata* is not yet adequately studied. Yet it is clear that it comprises diploid ($2n = 16$) forms (apparently the common chromosome level) as well as tetraploid ($2n = 32$) and hexaploid ($2n = 48$) chromosome races (Watkins 1981; Beridze and Kvatchadze 1981) which are morphologically very similar to their diploid relatives. Some cultivated clones are triploid.

Cultivated hexaploid *P. domestica* is thought to be a polyploid product of a cross between *diploid divaricata* (= *cerasifera*) plums and another *Prunus* species, namely the tetraploid blackthorn or sloe, *P. spinosa* L. (Crane and Lawrence 1952, p. 237). The latter is a wild thorny bush widely distributed over central and northern Europe and the cooler and wetter areas in western Asia. However, the available cytogenetic evidence in support of such polyploid origin of the 6× cultivated *domestica* plums is far from being convincing; and comparative morphology does not confirm this supposition. The *domestica* plums look strikingly similar to the *divaricata* forms. Both groups intergrade and apparently intercross with one another. In contrast, *P. spinosa*, with its small and very astringent fruits, is morphologically distinct and seems to be isolated reproductively from the *domestica-divaricata* plums. Hybrids between these two groups are sterile. The hypothesis that 6× *P. domestica* has a CC SS S_1S_1 genomic constitution, i.e contains two genomes (S, S_1) contributed by *P. spinosa*, does not comply with the above facts. As argued by Zohary (1992), it is unlikely that the sloe contributed two genomes (or even a single genome) to the cultivated 6× *domestica* plums. Very probably, all subsp. *domestica* and subsp. *insititia* cultivars evolved directy from wild *P. divaricata* (= *P. cerasifera*) stocks.

Plum stones appear in several Neolithic and Bronze Age sites in Italy, Switzerland, Austria, and Germany. They are variable in shape, yet they fall within the morphological range of present-day *divaricata* and *insititia* plums (Bertsch and Bertsch 1949; Körber-Grohne 1984). All these finds seem to represent fruit collected from the wild or use of trees consciously spared at the edges of cultivation. We know very little about the beginnings of plum domestication, but since its culture depends on grafting, this fruit tree was probably taken into cultivation together with apples and pears. The earliest records of plum planting and grafting are from Roman times.

Cherries: *Prunus avium* and *P. cerasus*

Cherries are characteristic fruits of the cooler and temperate parts of the Old World (Fogle 1975). Two species are grown: the diploid ($2n = 16$) sweet cherry *Prunus avium* L., and the tetraploid ($2n = 32$) sour or morello cherry, *P. cerasus* L.

The sweet cherry, *P. avium*, is a rather tall tree (up to 20 m high) with sweet, round, red-black berries. Most cultivars are self-incompatible and require cross-pollination in order to set fruits. Maintenance is by grafting.

The cultivated clones of the sweet cherry are closely related to a group of wild and feral forms which are widely distributed over temperate Europe, northern Turkey, Caucasia, and Transcaucasia (Meusel *et al.* 1965, Vol. 2, p. 227; Webb 1968). The ripe fruits of these spontaneous cherries are smaller than those of the domesticated varieties and usually reach only 10 mm in diameter. They show the same black-red colour of *P. avium* and have either a sweet or a bitterish taste, but are never acid. Also the stones of the wild forms resemble closely those of the cultivars but are usually smaller (7–9 mm in the wild as compared to 9–13 mm in most of the cultivated forms).

The sour cherry, *P. cerasus*, is a smaller tree, rarely exceeding 8 m in height, with bright red berries and a characteristic acid taste. Most cultivars are self-compatible. Spontaneous individuals of this tetraploid cherry occur in Turkey, the Balkan countries, and west and central Europe. But they are regarded as feral derivatives of the fruit crop which probably evolved under domestication as a result of a cross between cultivated *P. avium* and another *Prunus* species, the ground cherry, *P. fruticosa* Pallas. The latter is a wild shrub distributed over parts of central and eastern Europe and north-east Turkey. Its small fruits have a cherry-like taste but are too astringent to be palatable.

Cherries were apparently collected from the wild long before their cultivation. The numerous finds of Neolithic and Bronze Age stones in central European settlements (Bertsch and Bertsch 1949) seem to represent wild forms. Cultivated cherries make their appearance rather late. The earliest report of cherry cultivation appears in classical times. Pliny tells that Lucullus, in the first century BC introduced to Rome a superior cherry variety which he obtained in the Pontus region (White 1970, p. 498). Large quantities of cherry stones were found in both Roman and Medieval contexts in Germany as well as other central European countries (Baas 1979; Kroll 1980b).

Latecomers: apricot, peach, and quince

Three additional rosaceous fruits: apricot, *Armeniaca vulgaris* Lam. (syn. *Prunus armeniaca* L.): peach, *Persica vulgaris* Miller (syn. *Amygdalus persica* L.; *Prunus persica* (L.) Batsch) and quince, *Cydonia vulgaris* Pers., appear as horticultural elements in classical times; and the characteristic large stones of the first two fruits have been unearthed in several sites in the Mediterranean basin from Roman times on. The apricot, the quince, and the peach, however, are not native to this region and they apparently reached the Near East and Europe from the east only in Greek and Roman times.

Apricot grows wild in the Tien Shan mountain range in central Asia, in eastern Tibet, and in north China. It was apparently grown in China before the recorded time of its introduction to the Near East, and horticulturalists conclude that its place of origin is east Asia (Bailey and Hough 1975; Watkins 1981). But we are still totally ignorant about the early history of this fruit crop in central Asia; that is, whether apricot cultivation in this area preceded or came after the Chinese culture. The apricot was introduced into the Near East from Iran or Armenia around the first century BC (Hopf 1973c). A few hundred years later it is already a well-established fruit tree in Syria, Turkey, Greece, and Italy.

The peach apparently had a similar history. Its wild forms are known from the mountain areas of Tibet and western China and there are records pointing to peach cultivation in China as early as 2000 BC (Hesse 1975). The peach is reported to have reached Greece at about 300 BC from Persia; but the Romans did not cultivate it until the first century AD. Soon after, the peach is established as a valued fruit crop in the Mediterranean basin.

The quince, *Cydonia vulgaris* Pers., grows wild in Caucasia, north Iran, and the Kopet Dagh range. The cultivars are widely used in south-west Asia and in the Mediterranean basin. The fruits are not eaten raw but cooked or utilized to prepare jellies and beverages. There is no archaeobotanical documentation on this fruit crop. However, literary sources seem to indicate that also the quince reached the Mediterranean only in classical times.

Citron: *Citrus medica*

Citrus fruit trees (the genus *Citrus* L.) had their origin in south-east Asia and in India. The citron, *C. medica* L., is the only member of the citrus group which was already grown in the Near East in classical times. The fruit of the citron, with its characteristic thick rind, has been appreciated medicinally and used for preparation of citronate confection. The Jews adopted the citron to serve in the religious ceremony of the feast of the Tabernacle.

The citron very probably came to the Near East from India. Theophrastos' detailed description of its cultivation and propagation leaves little doubt that, by the end of the fourth century BC, *C. medica* was already well established in the east Mediterranean region. As Hjelmqvist (1979b) points out, some data suggest an even earlier arrival. Among them is his find of citrus seed in 1200 bc Hala Sultan Tekke in Cyprus.

Other species of citrus seem to have arrived in the Mediterranean basin only much later. The earliest among them seems to be the lemon, *C. limon* (L.) Burn. Lemon varieties with small fruits and thin rinds have been cultivated in the Near East since the Arab conquest. The principal commercial citruses of today, the orange and the tangerine, came to the Mediterranean basin only after the establishment of maritime trade between Europe and south-east Asia.

Numerous cultivars of citrus are apomictic. They set seed containing asexual embryos, which retain the genetic constitution of the mother plant. Such a mode of reproduction is exceptional among fruit crops and it allows the maintenance of desired genotypes by the simple planting of seed. Apomixis was apparently of major advantage in the early stages of domestication of citrus fruits. By this means vegetative propagation was possible before the invention of grafting.

Almond: *Amygdalus communis*

The cultivated almond, *Amygdalus communis* L. (syn. *Prunus amygdalus* Batsch., *P. dulcis* (Miller) D.A. Webb), is the most widely cultivated nut in the Mediterranean basin (Watkins 1981) and probably one of the early fruit-tree domesticants in Old World agriculture. Among the rosaceous fruit trees, almonds are the earliest to flower and set fruit. They thrive best in the relatively warm Mediterranean-type climates. Compared with olives and grapes, almonds can endure somewhat drier conditions. Most cultivated almonds, as well as the wild forms, are self-incompatible. Some cultivars, such as the Apulia or Bari group, carry a mutation that renders them self-compatible. The main trends of evolution under domestication were selection for types with sweet, non-poisonous seed, larger drupes, and softer, thinner shells. Almonds cannot be propagated by rooting and modern almond cultivation depends mainly on grafting. Yet, in contrast with most other fruit trees, almonds were and are also planted from seed. When sexually reproduced, seedlings vary considerably in the size and shape of their fruits and in the hardness and thickness of the shells. They also frequently segregate in seed bitterness. Most of the non-bitter individuals are very tasty. Since the non-bitter condition is governed by a single dominant mutation (Spiegel-Roy and Kochba 1981), 75 per cent or more of the progeny raised from seed harvested in non-bitter almond plantations produce sweet fruits. Seed planting necessitates rogueing of unwanted bitter individuals, as well as acceptance of wide variation in the shape and size of the fruits. Nevertheless, this has been a traditional way of almond cultivation in numerous localities in the Mediterranean basin.

Wild ancestry

The genus *Amygdalus* L. comprises 26 species distributed over south-west and central Asia and south-east Europe. The cultivated almond is closely related to an aggregate of wild forms native to the Levant countries. These wild almonds, placed taxonomically (Browicz and Zohary 1993) within *A. communis*, fall into two intergrading eco-geographical races:

   (i) Relatively large wild and weedy forms, *A. communis* subsp. *spontanea*, thriving in Mediterranean environments (350–800 mm annual rainfall) and resembling closely the cultivated varieties in flower and leaf morphology, in early blooming, and in their general habit and climatic requirements.

   (ii) More xeric, smaller wild forms, *A. communis* subsp. *microphylla* [= *A. korschinskyi* (Hand-Mazz.) Bornm.], occupying drier 'steppe forests' or steppe-like environments. These almonds have smaller leaves and fruits compared to their more robust Mediterranean counterparts.

Wild forms placed in subsp. *spontanea* are identified as the progenitor of the cultivated varieties (Browicz and Zohary 1993). They differ from the cultivated fruit tree mainly by their smaller fruits, harder shells with fewer pits and their intensely bitter seed. The bitterness represents a chemical defence. It is brought about by the presence of the glycoside amygdalin, which becomes transformed into deadly prussic acid (hydrogen cyanide) after crushing, chewing, or any other injury to the seed. The consumption of few dozen bitter seeds yields enough prussic acid to prove fatal for human beings. Domestication of the almond involved first of all the breakdown of the wild type bitterness and the selection of the non-bitter almonds for cultivation. Sweet cultivars, known as var. *dulcis*, contain only traces of amygdalin. Some of them are homozygous, others are heterozygous for the dominant mutation.

Like numerous other fruit trees, *A. communis* comprises a variable complex of interconnected wild forms, weedy derivatives, and cultivars in its main area of distribution. In the Levant, wild forms of subsp. *spontanea* grow in Mediterranean *maquis* and oak park-forest formations, thriving best on rocky slopes and escarpments facing south; while 'weedy' forms often colonize secondary habitats such as neglected plantations, edges of terrace cultivation, and roadsides. The latter are frequently feral derivatives, or products of spontaneous hybridization between cultivated and wild forms.

Several other *Amygdalus* species, growing outside the Levant, are also quite close to the crop and are interconnected to it by sporadic hybridization. Most prominent among them are: *A. webbi* Spach, native to the Aegean basin and south Italy; *A. fenzliana* (Fritsch) Lipsky in north-east Turkey and adjacent Caucasia; *A. kuramica* Korsh. in north-east Afghanistan; and *A. bucharica* Korsh. in Tadzhikistan and Uzbekistan. (For details on their distribution, see Browicz and Zieliński 1984.) All these wild almonds were very likely isolated from *A. communis* in the past. However, the spread of almond cultivation into their territories brought them in contact with the crop. As evident from spontaneous hybrids encountered in places of contact, this superimposition initiated hybridization. How extensive is the gene-flow between the cultivated crop and these wild almonds is hard to assess. Yet similar to the apple or to the pear, the presence of intermediates and recombinants indicates that introgression from the local wild species could have facilitated the development of locally adapted *A. communis* cultivars and/or helped in the establishment of local feral *A. communis* stands.

Archaeological evidence

The oldest almond remains found in archaeological excavations come from the Palaeolithic layers in the Franchthi Cave, southern Greece, and from Mesolithic and all Neolithic beds in this site (Hansen 1978, 1992), which shows that wild almonds have been in continuous use in this area.

Pieces of broken shell from Pre-Pottery Neolithic A Jericho were tentatively assigned to *Amygdalus* especially since some charcoal of *Amygdalus* wood was also found on the same level (Hopf 1983). Similar rare shell finds are available from early Neolithic Tell Aswad, Syria (van Zeist and Bakker-Heeres 1985). In later periods, almond remains come from late

Neolithic Dhali Argidhi, Cyprus (Stewart 1974), Çatal Hüyük, Turkey (Helbaek 1964a) Tepe Musiyan, Iran (Helbaek 1969) as well as Dimini (Kroll 1979), Sesklo (Kroll 1981a), and Sitagroi (Renfrew 1979, table 3) in northern Greece. They also occur in late Neolithic/Chalcolithic Hacilar, Turkey (Helbaek 1970).

Almonds are reported from Early Bronze Age beds of Bab edh-Dhra, the Dead Sea basin, Jordan (McCreery 1979). Here they appear with numerous remains of grape vine and olive and very probably represent cultivated trees. These could have been cultivated either under irrigation (near to the site) or under rain-dependent conditions (at higher elevations nearby). Remains of almond shells continue to appear in Bronze Age sites in both the Near East, Egypt, and Greece. They were found in *c.* 1325 BC Tutankhamun tomb in Egypt (Germer 1989a; Hepper 1990; see also Fig. 43). From classical times on, *A. communis* is recorded as a characteristic element of Mediterranean horticulture (White 1970, p. 259).

In conclusion, early archaeological remains of almonds are still scarce and they do not permit a clear distinction between wild and cultivated forms. Yet almonds appear to have been members of the earliest cultivated fruit-tree assemblage in the Old World. They were very probably taken into cultivation in the eastern part of the Mediterranean basin and more

Fig. 43. Dessicated almond fruits from Tutankhamun's tomb, Egypt. These fruits are now placed at the Royal Botanic Gardens, Kew, UK. (Photo: A. McRobb. With kind permission of F. Nigel Hepper, formerly Assistant Keeper of the Herbarium at Kew.)

or less at the same time that the olive, grape vine, and date palm were domesticated, i.e. not later than the 3rd millennium bc. Of clear advantage is the ability of the grower to raise attractive almonds from seed. Thus in spite of the fact that this plant does not lend itself to propagation from suckers or from cuttings, it could have been domesticated even before the invention of grafting.

Walnut: *Juglans regia*

The common or Persian walnut, *Juglans regia* L., is a traditional nut of Old World agriculture. It is a large tree which, in addition to its fruits, produces beautiful hard timber. The fruit is a drupe. A fleshy, green, outer layer encloses the 'nut' and usually dries and falls off at maturity. The hard shell of the nut represents the endocarp of the fruit; it encloses a single large, edible seed, which is rich in oil.

Juglans regia grows wild in mesic, temperate, deciduous forests of the Balkan, north Turkey, the south Caspian region, the Caucasus, and central Asia. It reappears in the Tien Shan province of western China. Wild trees produce variable small fruits (2–3 cm in diameter), which have relatively thick shells. Their seed is as tasty as that of the cultivated tree. The cultivated varieties have larger fruits which are 3–6 cm in diameter. They are grown today in the native distribution range of *J. regia* as well as in central and western Europe and in the more xeric environments in the Mediterranean basin and western Asia. In the latter, the walnut thrives best in cool, hilly areas and its cultivation usually includes supplementary irrigation in summer.

The walnut does not lend itself to rooting by cutting or to propagation by suckers. Cultivation depends mainly on grafting of selected clones, but, like almond, this nut tree can also be raised from seed, though by this method trees vary considerably in the size and shape of their fruit and in the thickness of the shells.

Information pertaining to the time and place of walnut domestication is still inadequate. *J. regia*, however, produces large quantities of wind-borne pollen which is easily recognized in pollen analysis. Palynological data are helpful in the reconstruction of the history of the walnut in Europe and western Asia. The pollen data (Bottema 1980; Bottema and Woldring 1984) indicate that *J. regia* disappeared from south-eastern Europe and south-western Turkey during the last glaciation (Würm glaciation), when it survived only in the Pontic (Black Sea) and the Hyrcanic (south Caspian) refugia. Walnut pollen reappears in the Balkans and in south-western Turkey very late, not before the middle of the 2nd millennium bc. This late appearance suggests that *J. regia* did not return to these areas as a post-glacial wild element but was reintroduced by humans. If this is true, the spontaneously growing walnuts in the Balkans and central Europe represent feral derivatives of cultivated walnuts introduced by humans as recently as the Bronze Age. This points to north-eastern Turkey, the Caucasus, and north Iran as the most plausible area of walnut domestication.

Chestnut: *Castanea sativa*

The sweet chestnut, *Castanea sativa* L. is a valued nut crop in the humid parts of the Mediterranean basin. Until recently chestnuts were utilized extensively in various south European countries and in Asia minor. The nuts formed an important part of the diet in the traditional farming communities and were also used as an animal feed. In recent years chestnut production in south Europe has suffered debilitating damage from the attacks of two fungal diseases recently introduced from America: the chestnut blight caused by *Endothia parasitica*, and the ink root disease brought about by *Phytophthora cambivora*. In many places chestnut stands were decimated and production fell to only 10–20 per cent of its level at the beginning of this century.

Chestnut cultivation is based primarily on the selection of clones producing large and tasty nuts and their maintainance by grafting. Wild or naturalized stands which thrive on steep slopes and acid soils are frequently grafted *in situ* with scions producing superior nuts.

Castanea L. comprises some 10–12 species spread over the northern hemisphere (Richardson 1981). The cultivated chestnut is closely related to a variable aggregate of wild and feral forms which occur in the northern parts of the Mediterranean basin, north Turkey, and Caucasia (for details, see Meusel *et al.* 1965, Vol. 2, p. 121; Browicz 1982, map 48). Such forms also penetrate deeply into the climatically milder areas in central and western Europe. Because of their close affinity with the cultivated clones botanists include these spontaneous chestnuts within *C. sativa*.

Our knowledge of the place of origin and time of domestication of *C. sativa* is still inadequate. Palynological data indicate that *C. sativa* disappeared from southern Europe during the Würm glaciation and apparently survived only in south-west Asian refugia. Chestnut pollen appears in quantity in Anatolia and Greece only at about 1500–1300 bc (van Zeist 1980). A similar increase in pollen frequency is found in Italy and other west Mediterranean sites but only at the start of classical times (van Zeist 1980). This strongly suggests that similar to the walnut (p. 177), *C. sativa*, did not arrive in western Turkey, Greece, and the western Mediterranean countries as a wild element but was introduced by humans. If this is true, the spontaneously growing chestnuts in Italy, southern France, Spain, and adjacent places in western and central Europe do not represent genuinely wild populations but naturalized elements derived from the cultivated chestnuts brought by humans. This again points to north Turkey and the

Caucasus as the most probable area for the initial chestnut domestication. However, as van Zeist (1991) remarks, chestnut remains were retrieved from eleventh century bc Greifensee-Bochen near Zürich; and its presence was attested (by charcoal remains) in *c*. 1500 bc Monte Leoni, north Italy. These finds suggest that *C. sativa* might have survived the last glaciation in some local refugia in Europe, and that its reoccupation might not necessarily have been exclusively from the east.

Hazels: *Corylus avellana* and *C. maxima*

Corylus L. is a genus of about 15 species of deciduous shrubs and trees distributed over the temperate parts of Europe, Asia, and north America (Wright 1981). The common hazel of Europe and west Asia is the European hazel, *C. avellana* L., a many-branched shrub or small tree. Its nuts, measuring 10–25 mm in length, are covered with a relatively short involucre, almost the size of the fruit. *C. avellana* is a common component of the broad-leaved oak and beech forest belt of temperate Europe, Caucasia, north Turkey, and the Caspian belt of Iran (for details of its distribution, consult Meusel *et al*. 1965, Vol. 2, p. 118; Browicz 1982, map 48).

   The tasty, rounded or ovoid nuts are easily collected and shell remains of *avellana* nuts have been repeatedly retrieved from many Neolithic, Bronze Age, Classical, and Medieval contexts all over Europe. The European hazel was also taken into cultivation and planted both for its nuts and for a supply of branches used for the preparation of hurdles and walking sticks. Superior clones were traditionally kept by the layering of branches. Like almonds, hazels were also grown from seed obtained from superior individuals. Although sexual reproduction results in a wide variation in nut size and shape, all nuts, even the small ones, are tasty. Where and when the domestication of *C. avellana* was started is not yet clear; but apparently this shrub was already planted by the Romans (White 1970, p. 259). In recent years the cultivation of *C. avellana* in west and central Europe underwent a sharp decline, as it was replaced by the true 'filbert' (*Corylus maxima*) or by modern hazel derivatives of hybrid origin.

   Corylus maxima Mill. is a hazel closely related to, fully interfertile with, and frequently included taxonomically within *C. avellana*. It is native to the Balkan, north Turkey, and Caucasia. The fruit is somewhat larger than in typical *C. avellana*; and it is enclosed within a tubular involucre constricted above the nut. Also *C. maxima* has very attractive nuts (Lagerstedt 1975) which were – and still are – extensively collected from the wild; and superior plants were taken into cultivation (as clones). Becasue of the bigger fruit size and better yields, planting of *maxima*-type clones spread considerably in recent times and often replaced *C. avellana* in west and central Europe. Today the world production of hazel nut is primarily based on *C. maxima* cultivars, which are called the true filberts, or on modern hybrid derivatives obtained from crosses between *C. maxima, C. avellana* and several additional *Corylus* species.

Pistachio: *Pistacia vera*

The pistachio nut, *Pistacia vera* L., ranks among the most drought resistant fruit trees of west Asia. It is extensively grown in south Turkey, Iran, and central Asia for its oval (*c.* 1–2 cm long), tasty nuts. The shell encloses a single oil-rich seed and splits longitudinally, along its lateral suture, when the nut is ripe. *Pistacia vera* is a dioecious, wind pollinated tree and fruit-bearing female clones have been traditionally planted intermixed with male individuals. Cultivation depends on grafting. Frequently the cultivars are grafted on stocks of other wild *Pistacia* species (*P. palaestina* or *P. atlantica*).

Pistacia vera grows wild in north-east Iran, north Afghanistan, and in the middle Asian republics – Uzbekistan, Tadzhikistan, Kirgizia and the southern most parts of Turkmenia and Kazakhstan, where this xeric tree constitutes a conspicuous element in the local 'steppe forest' vegetation belt. (For details on the distribution of the wild forms see Browicz 1988, map 5.) Wild forms have smaller (but edible) fruits which are collected and consumed by the local nomads. Occasionally, wild *P. vera* individuals are grafted *in situ* with cultivated clones.

The distribution area of the wild forms of *P. vera* indicates that this tree should have been brought into cultivation in central Asia. Indeed, the earliest archaeological documentation of this nut tree comes from the Kerman district, Iran. Two entire fruits of this tree were discovered in late Neolithic and Bronze Age Tepe Yahya (Costantini and Costantini-Biasini 1985). The total dependence of pistachio on grafting today suggests relatively late domestication. There are no signs of *P. vera* remains in the Near East before classical times, although a single *P. vera* half-shell was reported by Renfrew (1973) in late Neolithic Sesklo; and another single fruit was found by Kroll (1983) in Bronze Age Kastanas, Greece. In our view both these finds are questionable, they very probably represent more modern intrusions into the reported layers.

6
Vegetables and tubers

Compared to cereal grains and stones of fruits, vegetables and tuber crops are highly perishable, and their soft parts stand only a rare chance of charring and being preserved in archaeological contexts. It is therefore not surprising that very few remains of vegetables and tubers have been found in archaeological excavations. The only exception is Egypt, where numerous non-carbonized vegetables and tubers survived because of the extreme dryness. The available archaeobotanical data on the early phases of vegetable domestication are therefore embarrassingly fragmentary. Yet this is partly compensated by evidence from Mesopotamian Bronze Age literary sources and by drawings and descriptions found in Egyptian tombs. The combined evidence shows that at least since the start of the 2nd millennium bc, vegetable gardens constituted an integral element of food production both in Babylonia and in the Nile Valley. Furthermore, most of the plants grown at these early times are already satisfactorily identified: watermelon, melon, leek, garlic, onion, lettuce, and chufa appear to be major constituents of Bronze Age vegetable production.

Towards the end of the 1st millennium BC, the number of vegetable crops grown in south-west Asia, Egypt, and Europe seems to have increased considerably. Numerous additional vegetables are described in Greek, Roman, and Jewish classic sources (for reviews see Lenz 1859; White 1970; Körber-Grohne 1987). Prominent among them are: cabbage *Brassica oleracea* L., beet *Beta vulgaris* L., turnip *Brassica campestris* L., celery *Apium graveolens* L., carrot *Daucus carota* L., endive *Cichorium endiva* L., chicory *Cichorium intybus* L., globe artichoke *Cynara scolymus* L., and garden asparagus *Asparagus officinalis* L. Significantly, all these new cultigens have their wild relatives growing in south-west Asia and/or in Europe. All could have been taken into cultivation in this part of the world. The following paragraphs sum up the information on what seems to have been the first vegetable and tuber crops of the Old World.

Watermelon: *Citrullus lanatus*

The watermelon, *Citrullus lanatus* (Thunb.) Mats. & Nakai (syn. *C. vulgaris* Schrad.) was cultivated in the Nile Valley at least since the start of the 2nd millennium bc. Several finds of its characteristically large seed (as well as a single find of leaves) are reported by Keimer (1924, pp. 17–18), the oldest of them dating from the 12th dynasty (twentieth to eighteenth centuries bc). Numerous seeds were found in *c.* 1325 BC Tutankhamun tomb (Germer 1989b; Hepper 1990). Also the Agricultural Museum of Dokki, Cairo, has a sample of watermelon seed retrieved from the New Kingdom Thebes (Darby *et al.* 1977). More recently, van Zeist (1983) confirmed these early finds and reported on the presence of watermelon remains in the foundation deposits of the wall of an 18th dynasty temple near Semna, Nubia.

The cultivated watermelon is closely related to and fully interfertile with the wild Colocynth, *Citrullus colocynthis* (L.) Schrad., and the crop is very probably derived from this wild watermelon (Zohary 1983). In its general habit *C. colocynthis* resembles closely the cultivated crop but it is usually perennial and it bears relatively small (5–8 cm in diameter) fruits which have a spongy pulp and are intensely bitter. The colocynth is widely distributed over sandy areas in the deserts and semi-deserts of North Africa and west Asia, and its fruits are used for their strong purgative effect. They are still collected today by nomads for their medical value. The seed are sometimes used, after special preparation, as human food (Osborn 1968). Significantly, the characteristically small seed of *C. colocynthis* appears in several Egyptian and Near Eastern sites, including 3800 bc Nagada-Khattara (W. Wetterstrom, personal communication) and Pre-Pottery Neolithic B Nahal Hemar Cave Israel (Kisalev 1988), indicating that wild watermelon was very probably used by humans even prior to its domestication.

Melon: *Cucumis melo*

The cultivated melon, *Cucumis melo* L., is a second cucurbit which was probably brought into cultivation in south-west Asia or in Egypt at a relatively early date. This is a very variable crop which includes both (i) sweet fruited varieties (melons or muskmelons) and (ii) unsweet green-fruited forms (chate melons). The latter are rarer today, frequently bear bent fruits, and are consumed like cucumbers. They were traditionally referred to as *C. chate* Hasselqu., or *C. melo* L. var. *chate* (Hasselqu.) Naud.

The wild ancestry of the cultivated crop is already well established (Jeffrey 1980). The domesticated varieties show close morphological resemblance and full interfertility with a variable group of wild and weedy annual melons distributed over the subtropical and tropical parts of Asia and Africa. The more xeric wild type (frequently refered to as *Cucumis callosus* (Rottl.) Cong.), native to central Asia and the Near East, is closest to the cultivated melon forms raised in west Asia and the Mediterranean basin. The Indian and east Asiatic cultigens show closer affinities to a second, more tropical, wild melon race.

The archaeological remains of melons are few. Yet they seem to indicate that *C. melo* was cultivated already in the Bronze Age. Illustrations of offerings of what are clearly the bent fruits of the green-fruited melons (var. *chate*) decorate several ancient Egyptian tombs – from the Old Kingdom period on. (Keimer 1924; pp. 14–17). Three carbonized seeds are available from Late Bronze Age Tiryns, Greece (Kroll 1982) and few others from *c.* 2000 bc Shahr-i Sokhta, east Iran (Costantini 1977).

The melon, *C. melo* (úkuš in Sumerian and *giššû* in Akkadian), was the principal (and the oldest) cucurbitaceous vegetable of ancient Mesopotamia (at least from the beginning of the 2nd millennium bc on). The 'small' and 'finger' varieties mentioned by Stol (1987b) were very probably cucumber-like, non-sweet forms. The 'large' and the 'ripe' forms could represent more advanced non-sweet or even sweet-fruited melons. The sweet melons are the *melopepo* of the Greeks.

Finally, it is noteworthy that a third Old World principal cucurbit, namely the cucumber, *Cucumis sativus* L., does not belong to the native vegetable ensemble of the Near East and the Mediterranean basin. Wild forms of *C. sativus* occur only in the Himalayas and in adjacent territories east of this mountain belt (Jeffrey 1980). *C. sativus* was, therefore, probably taken into cultivation in northern India and arrived in the Mediterranean basin rather late. The first signs of *C. sativus* in the Near East come from *c.* 600 BC Nimrud, Iraq (Helbaek 1966b). But only two seeds were recovered. The cucumber was already known to the Greeks and Romans.

Leek: *Allium porrum*

The garden leeks and kurrats, grouped in *Allium porrum* L. (syn. *A. kurrat* Sfth. & Krause), seem to have ranked as one of the popular vegetables in the ancient Near East. It is apparent from ancient Egyptian references as well as from several finds of dried specimens (Keimer 1984; Täckholm and Drar 1954, p. 103), that *A. porrum* was part of Egyptian food production from at least the 2nd millennium bc onwards. Outside Egypt remains are few, but include either wild or cultivated material from Early and Middle Bronze Age Jericho (Hopf 1983). Linguistically leek is identified with Akkadian *Karašum* and Sumerian *Ga-Rash-Shar*. Available texts indicate that it was grown in Mesopotamia already at the beginning of the 2nd millennium bc (Stol 1987a).

Leek is obviously a Mediterranean or Near Eastern element and its wild progenitor is well known (McCollum 1976). The more robust varieties ('garden leek' or *A porrum* L. *sensu stricto*) and the slenderer leafy kurrats (traditionally called *A. kurrat* Sfth. & Krause), are all closely related to wild *Allium ampeloprasum* L., a wild leek which is widely distributed in the Mediterranean basin.

Garlic: *Allium sativum*

Garlic, *Allium sativum* L., also seems to be an early constituent of the Near East vegetable garden. In contrast to the leek, garlic is maintained

Fig. 44. Dessicated bulbs of garlic from Tutankhamun tomb, Egypt. (Photo: H. Barton. With kind permission of Griffith Institute, Ashmolean Museum, Oxford.)

by vegetative propagation. Egypt provides early archaeological evidence for this *Allium* crop. Excellently preserved garlic (Fig. 44) was found in *c.* 1325 BC Tutankhamun's tomb (Germer 1989a; Hepper 1990), and several well preserved dry remains of garlic are available from other 18th dynasty and later tombs. (Täckholm and Drar 1954, p. 102; Germer 1989b). Outside Egypt, early records are few, but a large number of carbonized garlic cloves were uncovered from the 2nd millennium bc Tell ed-Der, Iraq (van Zeist and Vynckier 1984). Also linguistically garlic is well recognized, and numerous cuneiform records indicate (Stol 1987a) that garlic was cultivated in Mesopotamia since at least the beginning of the 2nd millennium bc.

The wild ancestry of cultivated garlic has not so far been definitely established. The crop bears the closest morphological resemblance to *Allium longicuspis* Regel, a wild garlic distributed in central Asia, north Iran, and the south-eastern border of Turkey. This species is considered to be the most probable candidate for the ancestry of the cultivated clones (Stearn 1978). Accordingly, garlic could have been taken into cultivation somewhere on the north-east fringe of the Near East arc.

Onion: *Allium cepa*

A third member of the genus *Allium* which was apparently highly appreciated in the ancient Near East and the Mediterranean basin is the onion, *A. cepa* L. Wall carvings and drawings depicting offerings of onions and illustrations showing how onions are planted and watered appear in numerous Egyptian tombs from the Old Kingdom onwards (Germer 1989b). This includes Unas (*c.* 2420 bc) and Pepi II (*c.* 2200 bc) pyramids. Well-preserved onion remains are available from 18th-dynasty tombs in Egypt (Täckholm and Drar 1954, p. 104; Germer 1989b). They are complemented by somewhat later finds, including bulbs placed in mummies. Outside Egypt undisputed early archaeobotanical records of onions are not available. However, onion is linguistically well-studied; and similar to garlic, cuneiform sources indicate (Stol 1987a) that this vegetable was grown in Mesopotamia from the 2nd millennium bc on.

Like garlic, onion is not a Mediterranean element. Its progenitor is not yet convincingly identified. The crop shows close morphological affinities to *A. oschaninii* B. Fedtsch., which grows wild in north Iran, Afghanistan, and adjacent territories in central Asia (Stearn 1978). However, crosses between wild *A. oschaninii* and cultivated *A. cepa* gave rise to sterile hybrids (Hanelt 1985) indicating that this wild onion cannot be considered as the direct source from which the vegetable was derived.

Lettuce: *Lactuca sativa*

Illustrations of rosettes of a tall, large vegetable with subulate leaves appear in numerous Old Kingdom and Middle Kingdom tombs and monuments in Egypt (Keimer 1924; Körber-Grohne 1987, plates 81, 82). Most archaeobotanists agree that these leafy vegetables depict the garden lettuce, *Lactuca sativa* L. Unfortunately, the available illustrated record is not yet complemented in Egypt by dry remains of lettuce leaves or lettuce rosettes, although seed of lettuce have been reported (Germer 1985, p. 185). Outside Egypt indisputable records of *L. sativa* cultivation appear only in Greek and Roman literature.

Lettuce is definitely a west-Asiatic and Mediterranean element and its wild ancestor is already well recognized (Zohary 1991). The cultivated vegetable shows close affinities to a group of *Lactuca* species centred in west Asia. Its closest wild relative is weedy *L. serriola* L., which abounds in the Mediterranean basin and the Near East and also occupies (as a weed) vast areas outside these territories (including very successful post-Columbus colonization of the New World).

Chufa or rush nut: *Cyperus esculentus*

Chufa or rush nut, *Cyperus esculentus* L., ranks among the oldest cultivated plants in Egypt. It is a vegetatively propagated tuber crop, and its small (1.5–2.5 cm) ellipsoid-globose tubers appear in large quantities in Egyptian archaeological contexts from predynastic times on (Täckholm and Drar 1950). *C. esculentus* obviously represents a local domestication. The wild and weedy race of this sedge, var. *aureus* Richt., with its smaller and more fibrous tubers, grows today in the Nile Valley. While Chufa was no doubt an important food element in ancient Egypt throughout dynastic times, its cultivation in ancient times seems to have remained (totally or almost totally) an Egyptian speciality. There are almost no contemporary records of this tuber-bearing sedge from other parts of the Old World.

Cabbage: *Brassica oleracea*

Cabbage, *Brassica oleracea* L. is a cruciferous plant which was already well established as a garden vegetable during Greek and Roman times, on which we have no archaeological records. Wild cabbages thrive in the Mediterranean basin as well as along the Atlantic coast of Europe, and the plant should have been domesticated in this general area (Snogerup 1980). In classical times leafy, cauline cabbages prevailed. Theophrastus mentions three kinds of cabbage: curly leaved, smooth leaved, and wild type. Hearting cabbage, cauliflower, Brussel sprouts, kohlrabi, and other familiar varieties of this exceedingly variable vegetable are later derivatives.

Beet: *Beta vulgaris*

Beet, *Beta vulgaris* L., was a well-established vegetable in classical times. Because beet is linguistically well recognized, written records on this vegetable are valuable. The earliest documentation comes from eighth century BC Babylonia. Beet was grown in the garden of Merodachbaladan (Körber-Grohne 1987, p. 211). Greek, Roman, and Jewish literary sources provide clear information that already in the first century BC the crop was represented by several leafy forms (chards). Cultivars with swollen roots appeared later. Unfortunately, there are no archaeological records of *B. vulgaris* from pre-classical times. Consequently there is no answer to the question of when and where beet was taken into cultivation, although beet is obviously a Mediterranean element; the wild forms from which the crop could have been derived are widely distributed over the Mediterranean basin and the Near East (Ford-Lloyd and Williams 1975).

7

Condiments

Like the vegetables, several condiments native to west Asia and the Mediterranean basin seem to have entered cultivation rather early, and the tradition of their planting and usage is already well documented in classical Greek, Roman, and Jewish sources. Very few remains of this group of crops have been discovered in archaeological contexts, and our knowledge of the beginning of their domestication is consequently insufficient. The following paragraphs review the available evidence on what seem to have been the first condiments of the Old World. The mustards have already been surveyed in Chapter 5.

Coriander: *Coriandrum sativum*

Coriander, *Coriandrum sativum* L., an annual umbellifer cultivated for its aromatic fruits, was well known in classical times. Plant remains and linguistic evidence indicate that its use started much earlier. Fifteen desiccated coriander fruits were retrieved from Pre-Pottery Neolithic B Nahal Hemar cave, Israel (Kislev 1988). If these fruits are not intrusive, they represent the earliest archaeological find of this condiment yet. A large sample of the round small fruits of *C. sativum* was found in *c.* 1325 BC Tutankhamun tomb (Germer 1989a; Hepper 1990). It is also present in a 21st-dynasty (*c.* 1000 BC) tomb in Deir-el-Bahari. Additional finds are available from later sites in Egypt (for review, see Darby *et al.* 1977, p. 798). Coriander remains were also discovered in the 2nd millennium bc Tell ed-Dar, Syria (van Zeist and Vynckler 1984), in Iron Age Deir Alla, Jordan (Neef 1989), and in Late Assyrian Nimrud (Helbaek 1966b). A single half-fruit of *C. sativum* was found in Late Bronze Age Apliki, Cyprus (Renfrew 1973, p. 171).

Today *C. sativum* occurs spontaneously over wide areas of Old World agriculture and it is hard to define exactly where this plant is wild and where it only recently established itself as a weed or as a naturalized element. In the Near East and the east Mediterranean basin *C. sativum* seems to grow wild in oak park-forests, oak scrub, and adjacent steppe-like formations and it is probable that these territories were the source of both the cultivated forms and the weedy races.

Cumin: *Cuminum cyminum* and dill: *Anethum graveolens*

Two additional aromatic umbellifers: cumin, *Cuminum cyminum* L., and dill, *Anethum graveolens* L., were part of Greek and Roman agriculture, and we also have some information on earlier phases of their domestication. Cumin seeds were uncovered in the 2nd millennium bc Tell ed-Dar, Syria (van Zeist and Vynckler 1984) as well as in Iron Age Deir Alla, Jordan (Neef 1989). Cumin is also represented by few seed discovered in New Kingdom Deir el Medineh in Egypt (Germer 1985, pp. 135–6). Several twigs of dill were found in the tomb of Amonophis II 18th dynasty, *c.* 1400 BC (Germer 1985, pp. 142–3).

Wild forms of cumin are unknown in the Near East but occur in central Asia. Wild and weedy types of dill are widespread in the Mediterranean basin and in West Asia.

Black cumin: *Nigella sativa*

Black cumin, *Nigella sativa* L., a member of the buttercup family Ranunculaceae, was another traditional condiment of the Old World during classical times and until recently its black seeds were extensively used to flavour bread. Archaeological information on the early cultivation of black cumin is still scanty, but seeds of *N. sativa* were reported from ancient Egypt (Braun 1879), including *c.* 1325 BC Tutankhamun tomb (Hepper 1989; Germer 1989a). Linguistic considerations suggest that this plant was also grown in ancient Mesopotamia (Thompson 1949).

Wild forms of *N. sativa* occur in south Turkey, Syria, north Iraq, as well as in several adjacent territories, which implicates the Near East as the place of domestication of this condiment.

Saffron: *Crocus sativus*

Saffron, *Crocus sativus* L., of the iris family Iridaceae, is another member of the early group of condiments. The long, scarlet style lobes harvested from its flowers were highly valued for flavouring foods and for colouring them golden-yellow. They were also used for dying textiles. Saffron was already extensively grown in the Near East and the Mediterranean basin in classical times, and it maintained its role as an attractive dye until the beginning of this century. Recently, production became increasingly impractical because the collection of the styles requires a vast amount of labour, and today saffron cultivation is fastly disappearing and this *Crocus* survives only as a relic crop.

No remains of *C. sativus* have yet been traced in archaeological excavations and indeed the chances for the preservation of the delicate stigmas seems rather low. However, Minoan (1600 bc) pottery and frescoes depict *Crocus* flowers with long exserted red stigmatic branches (Mathew 1977), and these very probably represent *C. sativus*. Linguistic evidence also suggests that saffron was brought into cultivation long before classical times.

The cultivated saffron is an autumn flowering sterile triploid ($3\times = 24$ chromosomes) maintained by vegetative propagation. It is morphologically closest to the Greek species *C. cartwrightianus* Herbert and is possibly a clonal selection of this smaller diploid ($2n = 16$) species (Mathew 1977).

8

Dye crops

A group of dye plants, native to south-west Asia and the Mediterranean basin seems to have entered cultivation already prior to classical times; and they are amply documented in Greek, Roman, and Jewish literary sources. Woad *Isatis tinctoria* L., dyer's rocket *Reseda luteola* L., Madder *Rubia tinctorum* L., safflower *Carthamus tinctorius* L., and indigo *Indigofera tinctoria* L. were characteristic dye crops of Old World agriculture, and their dye stuffs were extensively used to colour textiles and leather. They continued to be highly appreciated crops until the end of the nineteenth century when the invention of low priced synthetic dyes caused their collapse. In a matter of a few decades the traditional dye crops of the Old World became commercially redundant and their cultivation died out. Today, they represent agricultural drop-outs. Most of the cultivars are extinct. Only few forms survive in botanical gardens or in the hands of hobbyists.

To date only few remnants of dye crops have been uncovered in archaeological excavations. Consequently, we know very little about the beginnings of dye plant cultivation. Yet early literary sources, together with some analyses of dye elements colouring ancient Egyptian textiles (Eastwood 1984; Germer 1992) make it clear that already in the 2nd millennium bc dye crops were extensively used. The following paragraphs review the available evidence on cultivation of dye plants in the Old World.

Woad: *Isatis tinctoria*

Woad, a member of the mustard family Cruciferae, was a principal dye crop in Europe and in south-west Asia. Until the end of the nineteenth century woad was extensively cultivated (by seed planting) for its blue indigo dye, which was used to colour textiles. The pigment was extracted from the leaves of the vegetative parts (rosettes) of the plants. To obtain the dye the leaves were dried, powdered, and fermented – a tedious and very smelly process. Today natural indigo is a commerical rarity. In the textile industry it was replaced by synthetic aniline dyes.

Woad *Isatis tinctoria* L. is usually a biennial plant with characteristic pale yellow flowers and winged fruits. It is a native of south-west Asia and the Aegean area. It also extends to temperate Europe. Some of the present weedy populations of *I. tinctoria* (particularly those of temperate Europe) may be feral derivatives of the former crop.

Similar to madder, (see below), classical literary sources show that already in the 1st millennium BC woad was extensively used in south-west Asia, in the Mediterranean basin, and in temperate Europe (Körber-Grohne 1987). In addition, indigo dye (indigotin) was detected in several blue-coloured ancient Egyptian textiles found in 18th dynasty (1370 bc) Tell el 'Amarna, Egypt (Eastwood 1984; Germer 1992). While it is impossible to distinguish chemically between indigo extracted from *Isatis* or from *Indigofera* (p. 193), only the first seems to have been present in the Mediterranean basin in classical times. *Indigofera* arrived much later.

Dyer's rocket: *Reseda luteola*

Dyer's rocket or weld *Reseda luteola* L. was the source of a flavon-type yellow pigment extracted from its roots. Also this plant was extensively cultivated (by seed planting) until the beginning of this century when the natural dye was replaced by cheaper synthetic yellow dyes.

Dyer's rocket is an annual or a biennial plant with erect stems and with yellow inflorescences. The plant is very likely indigenous to the east Mediterranean basin and to south-west Asia, and seems to have become naturalized far beyond its original native range.

Classical literary sources show that dyer's rocket was cultivated and extensively used in the Old World already in the 1st millennium BC (Körber-Grohne 1987, p. 417). Similar to woad and madder it was probably taken into cultivation much earlier.

Madder: *Rubia tinctorum*

Madder was, until the end of the nineteenth century, one of the most appreciated dye plants of south-west Asia and Europe. It was extensively grown (by vegetative propagation) all over this area for its rhizomes from which a brilliant red pigment (alizarin) was extracted. Alizarin was widely used to colour linen, wool, cotton, and leather. The dye gets fixed to the textile fibres only after their treatment with a mordant (alum salts). Furthermore, mordanting with different metals produces different hues (aluminium alum induces dark red coloration; iron alum results in brown-red colour; chromium alum produces in red-violet colour).

Pliny (Natural History XXXV xlii) described eloquently the technique of madder dyeing in Egypt:

They employ a very remarkable process of colouring textiles. After rubbing the cloth, which is white at first, they saturate it not with the dye – but with mordants that are calculated to absorb the colour. This done, the textiles still unchanged in their appearance are plunged into a cauldron of boiling dye and are removed the next minute fully coloured. It is a remarkable fact, too, that although the cauldron contains a uniform dye, the material taken out is of various colours – according to the nature of the mordants that have been respectively added to the cloth. These colours will never wash out.

Madder *Rubia tinctorum* L. is a perennial herb with characteristic whorls of lanceolate leaves and climbing or staggering scabrous stems, that can reach the length of 1–2 m. Its wild forms are native to south-west Asia and central Asia (F. Ehrendorfer, personal communication). Spontaneous populations of *R. tinctorum* thrive also in the Mediterranean basin mostly in hedgerows, waste ground, and margins of cultivated fields. Very possibly they represent naturalized derivatives of former cultivation.

It is still impossible to conclude where and when madder was taken into cultivation or when the alizarin colouring technology was developed. Chemical analysis has detected madder dye in red-coloured flax textiles retrieved from 18th dynasty (1370 BC) Tell el 'Amarna, Egypt (Eastwood 1984; Germer 1992). Various Greek, Roman, and Jewish literary sources show that at the end of the 1st millennium bc *R. tinctorum* was extensively cultivated in Persia, Anatolia, and the Mediterranean basin (Körber-Grohne 1987, p. 420). Evidently, domestication of madder should have started long before these times.

Synthetic alizarin was invented in Germany in 1869. In a matter of only a few decades the low-cost synthetic equivalent replaced the much more expensive natural dye stuff and caused a collapse of madder cultivation.

True indigo: *Indigofera tinctoria*

In India, as well as in several other parts of south Asia and in Africa south of the Sahara, the traditional source for indigo dye was not woad but several members of the genus *Indigofera* (Leguminosae). A common cultigen was *I. tinctoria* L. Wild forms of this dye plant occur in India, and probably this species was domesticated in this subcontinent (Lemmens and Wessel-Riemens 1991). Several other species of *Indigofera* (some of them native to Africa) were also utilized. We do not know when *Indigofera* was first introduced to the Near East; but certainly it is a newcomer. Sound documentation on its cultivation is available from Egypt only from after the Arab conquest. It could have reached the warmer parts of the Near East somewhat earlier.

Safflower: *Carthamus tinctorius*

Safflower, *Carthamus tinctorius* L., of the sunflower family Compositae, was a traditional dye plant of the Old World. Its yellow-red flowers were extensively used to dye textiles and to colour foods. With the recent development of synthetic dyes the importance of this crop declined considerably, although since the 1950s safflower acquired a new use and has emerged as a modern oil crop.

Seeds of safflower were found in *c*. 1325 BC Tutankhamun tomb in Egypt (Germer 1989a; Hepper 1990) and well-preserved garlands made of *C. tinctorius* flowers were found adorning 18th-dynasty (middle 2nd millennium BC) mummies in Egypt (Keimer 1924; Täckholm 1976). Chemical analysis of Egyptian textiles dated from the 12th dynasty showed safflower to be one of the dyes used (Darby *et al.* 1977). These records indicate that *C. tinctorius* was well known in ancient Egypt. Outside Egypt the early history of safflower is much less known. A find of a single safflower seed comes from Middle Bronze Age 2000–1600 bc Feudvar, Yugoslavia (Kroll 1990), and the plant is well identified in early Mesopotamian cuneiform records, but the early history of safflower outside Egypt is still unknown.

The wild stock from which the domesticant could have been derived is well identified (for reviews, see Hanelt 1963; Knowles 1969). The cultivated *C. tinctorius* is interfertile and chromosomally homologous ($2n = 24$) with *C. persicus* Willd., an early summer thistle distributed over the Syrian desert, Anatolia, and Upper Mesopotamia; as well as the highly spiny *C. oxyacanthus* M. Bieb. which replaces the former in Iran, Afghanistan, and central Asia. This relationship suggests a Near Eastern or central Asiatic origin and domestication that must have preceded the Egyptian 2nd millennium bc archaeological finds.

9
Fruit collected from the wild

Side by side with remains of cultivated plants, archaeological sites frequently abound with remnants of *wild* plants. Some come from unwanted weeds and man-followers, but the bulk of this material usually represents intentional collection from the wild. Even today wild 'useful plants' constitute a regular food source in numerous peasant communities. Moreover, specific plants are highly appreciated as a substitute 'hunger food' and are extensively used when crops fail. Other species are collected for medicinal purposes or used as stimulants. Still other plants constitute fibre sources or are utilized for spicing, dyeing, or tanning. In earlier times such exploitation of wild sources was no doubt even more extensive. Every ecosystem and every major plant formation had their characteristic assemblage of useful wild plants.

Seed, hard shells, pips, and stones constitute the bulk of the archaeological evidence. Soft plants are rarely preserved, so the information on past collection of useful wild plants is partial. We know, for instance, almost nothing about the use of wild pot-herbs, salad elements, or tubers and bulbs in Neolithic and Bronze Age times. In contrast, we have rich documentation on wild plants collected for their seeds, berries, or nuts. The large quantity of wild plant material discovered in excavations is indeed impressive, and indicates that long after the firm establishment of agricultural practices, collection from the wild continued to supplement food production on a substantial scale. In most ancient cultures wild plants seem to have provided a major parallel source of food.

This chapter surveys briefly some of the European and west Asiatic wild food plants, the remains of which abound in archaeological contexts. Emphasis is given to trees and shrubs that were apparently not only utilized, but also partly protected and occasionally even planted, by humans.

Oaks: *Quercus*

Numerous species of oaks (the genus *Quercus* L.) grow as common constituents of major vegetation formations in the Old World, both in the temperate regions and in the areas with Mediterranean-type climates. The evergreen *Q. coccifera* L. and *Q. ilex* L. are typical of the Mediterranean vegetation, *Q. robur* L. and *Q. petraea* (Matt.) Liebl. are widespread in temperate Europe, and *Q. brantii* Lindley and *Q. ithaburensis* Decne. are the principal builders of the Near East oak park-forest formation. Oaks vary considerably in their ecological adaptations and in their shape and foliage, but all produce numerous acorns at the end of the summer. The size and the shape of the fruit vary considerably from species to species. Oaks in temperate areas tend to have small acorns, while oak species adapted to summer-dry conditions usually produce bigger fruits; indeed some of the Near East oaks (e.g. *Quercus ithaburensis* or *Q. brantii*) have very large acorns which may reach 5–6 cm in length and 2–2.5 cm in diameter. All oaks share two features: their acorns are rich in reserve materials (starch and oil), and the mature acorns contain an appreciable amount of tannins, which render them bitter and unpalatable. There is, however, considerable individual variation in the intensity of bitterness. In several oak species rare, exceptional, sweet-fruited indiviuals occur. Such trees are recognized and appreciated by local people. (For details of the distribution of the main oak species consult Meusel *et al.* 1965, and Browicz 1982.)

In traditional peasant communities in Europe and the Near East acorns were exploited in the autumn, normally as a supplement feed for domestic animals. Yet in times of famine, acorns were collected and consumed by humans after grinding and leaching or after roasting or boiling. The time of acorn maturation was critical. Cereals mature in the summer while oaks produce their fruits in the autumn. Farmers therefore knew when their grain crops failed and resorted to acorn collection as a 'bread of hunger'. In fact, the Kurds of the Zagros Mountains in Iran and Iraq still use this term when referring to the conspicuously large acorns of *Q. brantii*, the dominant oak species in this area.

Carbonized remains of acorns appear frequently in Neolithic and Bronze Age contexts in the Near East, Mediterranean basin and central and north Europe (see, for example, Jørgensen 1977). They represent collections from the wild. There are no signs of oak domestication; that is, the planting of 'sweet' clones. One of the reasons for that could be that the various *Quercus* species do not lend themselves to simple vegetative propagation.

Beech: *Fagus*

Two species of beech (the genus *Fagus* L.) are widely distributed over the temperate parts of Europe and west Asia where they frequently thrive as dominant or codominant elements in the main forest formations. The European beech, *Fagus sylvatica* L., is a principal forest tree of temperate Europe. The closely related oriental beech, *F. orientalis* Lipsky, plays a similar role in the humid parts of north Turkey, Caucasia, the Caspian belt of Iran, and north Greece. Both species are large, deciduous trees that bear rather small (1–2 cm long) triangular nuts enclosed (usually in groups of three) in spiny cupules; they are rich in tasty oil. (For details of their distribution, see Meusel *et al.* 1965, Vol. 2, p. 120.)

In traditional farming communities situated in beech forest areas, the ripe nuts were frequently collected in the autumn. They served as a source of valuable, edible oil, and pigs were taken to the woods to feed on the nuts. Collecting beech nuts is apparently a very old tradition. Remains of *Fagus* nut shells and/or cupules have been retrieved from Neolithic and younger sites in temperate Europe (for review see Hopf 1976).

Wild pistachio: *Pistacia atlantica*

Carbonized remains of the small nuts of wild pistachios, particularly of *Pistacia atlantica* Desf., appear repeatedly in Neolithic and Bronze Age contexts in the Near East, and it seems that the ripe fruits of this wild tree were extensively collected in these periods. *P. atlantica* is a common arboreal constituent of the oak park-forest belt of the Near East arc. It is one of the more xeric elements of this vegetation formation. Together with the wild almond it abounds at the drier boundary of the park forest facing the more arid steppes.

The fruits of *P. atlantica* are still collected today in the Near East and are occasionally sold in the old-type markets. There is no sign that *P. atlantica* was ever domesticated, but another species of this genus, *P. vera* L., was brought into cultivation in central Asia (see p. 180).

Hawthorns: *Crataegus*

Several species of hawthorns, *Crataegus* L., of the rose family Rosaceae, are common all over Europe and west Asia. They usually attain the size of tall shrubs or small trees and bear small pomes which contain two to five nutlets or stones. Most common in Europe are *Crataegus oxyacantha* L. and *C. monogyna* Jacq. The latter extends also into the mesic parts of south-west Asia. *C. aronia* (L.) Bosc. (including *C. azarolus* L.) and *C. pubescens* (C. Presl.) C. Presl. are the prominent hawthorns of the Near East. (For details of their distribution see Browicz and Zieliński 1984.)

In several parts of Europe the dry flowery pulp of the relatively small fruits of *C. oxyacantha* and *C. monogyna* has traditionally been eaten or added to flour. In the Near East, the more fleshy and tasty fruits of *C. aronia* and *C. azarolus* are extensively collected from the wild and eaten fresh. They are still brought to provincial markets. Both species show a wide variation in fruit quality. Superior individuals are often spared. Occasionally, seed from such trees are even planted at edges of cultivation or on non-arable lands.

Stones of *C. oxyacantha* and *C. monogyna* were discovered in several archaeological excavations in Europe extending from the Neolithic to Roman times (for enumeration of records see Hopf 1976, p. 137). The nutlets of *C. aronia* (reported as *C. azarolus*) were found by Helbaek

(1958a) in Bronze Age Lachish, Israel. They appear also in several other Near-Eastern sites.

Hackberry: *Celtis australis*

The Mediterranean hackberry or nettle tree, *Celtis australis* L., of the elm family Ulmaceae, is a large deciduous tree distributed over the Mediterranean basin (for details on its distribution, see Meusel *et al* 1965, Vol. 2, p. 123; Browicz 1982, map 93). It is valued both for its timber and for its globose fruits. The ripe fruits (9–12 mm in diameter) contain a large central stone covered with thin pulp and dark brownish skin. The pulp is edible and in traditional farming communities the ripe fruits are frequently collected. Occasionally *C. australis* is even planted (from seed), but usually on non-cultivated grounds.

Carbonized stones of *C. australis* were found in several Neolithic and Bronze Age sites in the Mediterranean basin such as Çatal Hüyük (Helbaek 1964) and Hacilar (Helbaek 1970) in Turkey. They indicate that the fruit of this tree were collected and utilized throughout the agricultural history of this region.

Celtis is a relatively large genus and several other species of hackberry in Asia, America, and Africa also have edible drupes which are valued by humans.

Christ's thorn: *Zizyphus spina-christi*

Christ's thorn, *Zizyphus spina-christi* (L.) Desf., a member of the family Rhamnaceae, is a spiny evergreen tree widely distributed over the savannah and thornbush areas of east Africa and south Arabia as well as in desert oases and warm, rain-fed territories in the Near East (M. Zohary 1973). The tree is common in Egypt, in the Jordan valley, and in the coastal plain of Israel and Lebanon. It bears round, cherry-sized, yellow fruits each containing a central stone coated by some edible pulp. The fruits have a long tradition of utilization, and are collected by bedouin in desert oases and by poor peasants who occasionally refrain from cutting down individuals with relatively tasty fruits, keeping them for both shade and food.

Remains of *Z. spina-christi* appear repeatedly in Egyptian archaeological contexts from pre-dynastic times onwards (van Zeist 1983; Wetterstrom 1984). The stones were also discovered in several Near Eastern Neolithic and Bronze Age sites indicating that this wild tree has been in use thoughout the agricultural history of these regions.

Another species of *Zisyphus*, namely the jujuba, *Z. jujuba* Mill., grows wild in south-east and south Asia. It is planted today as a minor fruit tree in China, the Indian subcontinent, central Asia, and the Near East.

Where and when jujuba was taken into cultivation is still hard to say but most authors regard India or China as the most likely places of its domestication. Stones of *Z. jujuba* were discovered in 5th millennium bc Mehrgarh (Jarrige and Meadow 1980) and 2nd millennium bc Pirak (Costantini 1979), Pakistan. They also appear in several Harappa culture sites. All these finds probably represent collections from the wild.

Raspberries and blackberries: *Rubus*

Members of the genus *Rubus* L. of the rose family, Rosaceae, are widely distributed over temperate parts of Europe, Asia, and America. The fruits are berries composed of numerous single-seeded druplets which are set together on a conical or rounded receptacle or torus. Most species of *Rubus* bear edible berries which attracted humans since early times. One of the most widespread and tasty wild species in Europe and west Asia is the diploid ($2n = 14$) red raspberry, *R. idaeus* L. This species has a circumboreal distribution (for details on its area, see Meusel *et al.* 1965, Vol. 2, p. 211) and is represented in east Asia and north America by additional distinct geographical subspecies. Several other members of the genus *Rubus* with attractive fruits are also common in Europe and west Asia, including the blackberries, *R. fruticosus* L. and *R. tomentosus* Borkh., the dewberry, *R. caesius* L., and *R. saxatilis* L. (For details on their distribution, see Meusel *et al.* 1965, Vol. 2, pp. 211–13).

The genus *Rubus* is exceptionally rich in species and forms, and frequently present a nightmare to taxonomists engaged in species delimitation (over 300 species have been described in Europe alone!). The genetic system of this genus holds the key for this complexity. *Rubus* does not possess a simple sexual reproductive system. In several types we find almost complete sexual reproduction, while in numerous polyploid, derived types there is almost full apomyxis (production of asexual seeds). In other words, numerous apomictic forms (clones) evolved in *Rubus* through hybridization and fusion of the variation present in several sexual, diploid ($2n = 14$) parental stocks. This resulted in the complex patterns of variation characteristic to this genus today.

Domestication of *Rubus* began relatively late. It started in Europe when the raspberry, *R. idaeus* L., was taken into cultivation in medieval monasteries, but distinct cultivars of raspberry were not recorded until late in the eighteenth century (Jennings 1976). At that time cultivated *R. idaeus* had already been introduced to North America and it was supplemented there by selections from the native blackberry *R. occidentalis*, as well as by spontaneous hybrids between the two species. More recent developments were hybrids between the European *R. ideaeus* and the local American subspecies of raspberry. The modern cultivated blackberries were developed by crossing several west American, east American, and European *Rubus* species.

Pips of several species of *Rubus*, with their characteristic reticulate surface sculpture, appear frequently in European archaeological contexts from Neolithic to Medieval times (for enumeration of the records, see Hopf 1976a, p. 137). The kidney-shaped seeds of *R. idaeus* are 2.2–1.7 mm long and those of *R. fruticosus* and *R. caesius* are more rounded and somewhat larger, attaining 2.9–1.7 mm in length.

Strawberries: *Fragaria*

Several species of the genus *Fragaria* L. of the rose family Rosaceae, grow wild in the temperate parts of Europe and south-west Asia. The strawberry 'fruit' is actually a 'false fruit', consisting of a swollen juicy receptacle on which the small true fruits or achenes are embedded. The small (up to 1 cm), tasty fruits of wild *Fragaria* species were collected from the wild in prehistoric and historic times. The most common wild strawberry in central and north Europe is the diploid (2n = 14) wood strawberry, *Fragaria vesca* L., which is widely distributed also in temperate Asia and North America (Meusel *et al.* 1965, Vol. 2, p. 218). The Polunitza strawberry, *F. viridis* Duch. (2n = 14) thrives in temperate Europe in meadows and along the edges of forests. It also extends to some parts of the Mediterranean region. The hexaploid (2n = 42) Musk or Hautbois strawberry, *F. moschata* Duch., occurs in the cooler and elevated parts of southern Europe, extending eastwards to Russia and Siberia.

The cultivation of strawberries is fairly recent. Attempts to domesticate the local strawberries started in Europe in the fourteenth and fifteenth centuries and their cultivation continued until the nineteenth century when they were almost totally replaced by the Pineapple strawberry, *F. ananassa* Duch. This octaploid (2n = 56) crop was developed in west Europe from a hybrid between two introduced American species, *F. chiloensis* Duch. and *F. virginiana* Duch., in the middle of the eighteenth century.

The small (0.8–1.5 mm long), roughly ovoid achenes or 'seed' of wild strawberries have repeatedly been retrieved from Neolithic, Bronze Age, Iron Age, and Roman sites in central and northern Europe (for enumeration of these finds, see Hopf 1976, p. 137). They indicate continued use of these plants since early prehistoric times.

Elders: *Sambucus*

The common elder, *Sambucus nigra* L., a member of the family Caprifoliaceae, is a deciduous shrub or small tree widespread over temperate Europe. It is characterized by large, flat, white-flowered umbels which later bear numerous reddish-black berries. The use of the elder is an ongoing tradition in Europe, where tea is made from the dry flowers, and the juicy fruits are used to prepare syrups and to ferment elderberry wine.

The flat, elliptical (*c.* 4 mm long) seeds of *S. nigra* appear in numerous prehistoric sites (for a list of finds, see Hopf 1976, p. 137), indicating that elderberries, together with blackberries (p. 199) and strawberries (p. 200) were extensively collected in ancient Europe.

Another species of elder, namely Dane's elder, *S. ebulus* L., grows in the warmer parts of temperate Europe. This is a smaller, herbaceous plant. The seeds of this elder also occur in European archaeological contexts. They are smaller (2.8 mm long) and rounder than those of *S. nigra*. The berries were also used for dyeing.

Cornelian cherry: *Cornus mas*

Cornus mas L. of the dogwood family Cornaceae, is a shrub or small tree distributed over south and central Europe, the Black Sea basin and the Caucasus. It bears rather large (16 × 7 mm) elliptical fruits each containing a single stone enclosed by some sweet pulp. The ripe, scarlet fruits are collected from the wild and utilized in several ways: for preparation of sherberts and jellies, and for fermentation of alcoholic drinks.

Stones of the Cornelian cherry were discovered in numerous Neolithic and Bronze Age sites in south and south-east Europe (for a list of finds, see Hopf 1976, p. 137). These records indicate a common prehistoric use of this wild shrub.

10

Plant remains in representative archaeological sites

This chapter summarizes the information on plant remains retrieved from representative Neolithic and Bronze Age sites in west Asia, Europe, and the Nile Valley. It presents a selected list which was complied in order to answer the question: if one has to sketch the origin and the early spread of cultivated plants in the Old World, what would be the minimum number of archaeological locations that could provide an adequate account on the present state of our knowledge?

The information is arranged country by country. Representative sites had to be selected from countries which are still very poorly studied, as well as from areas that had a long tradition of archaeological excavation. Numerous well analysed locations in intensively researched countries were, therefore, not included in this chapter, while poorer sites in less thoroughly surveyed countries do appear on the list. Altogether 150 sites (or groups of sites; and possibly of long continuity) were chosen to present the archaeological evidence as it stands today.

Most sites are radiocarbon-dated (uncalibrated). Traditional dating was given where radiocarbon analyses were not available. A chronological chart for the different regions is given on p.238.

For information on the geographic location of the sites consult Maps 21–25 (pages 240–4) and the general references given for each country. Large dots on the maps indicate a group of contemporary sites in the same area.

Iran

(General reference: Miller 1991.)

1. **Ali Kosh**, Deh Luran Plain, Khuzistan (Helbaek 1969). (i) Bus Mordeh phase (*c.* 7000 bc). Rich remains: brittle einkorn wheat (few); einkorn wheat (rare); emmer wheat (prevailing); brittle two-rowed barley (frequent); naked barley (few). Wild: *Linum* (rare); *Prosopis farcta, Pistacia atlantica* and *Capparis spinosa* (rare). (ii) Ali Kosh phase (6340 bc). Rich remains: brittle einkorn wheat (rare); einkorn wheat (rare); emmer wheat (prevailing); two rowed barley (few); naked barley (rare); lentil (rare). Wild: *Linum* (rare); *Avena* (few); *Prosopis farcta* (frequent);

Pistacia atlantica and *Capparis spinosa* (rare). (iii) Mohammed Jaffar phase (*c*. 5600 bc). Rich remains: emmer wheat (frequent); two-rowed barley (prevailing); six-rowed barley (rare); lentil (rare). Wild: *Linum* (rare); *Avena* (frequent); *Pistacia atlantica* and *Capparis spinosa* (rare).

2. **Tepe Sabz**, Deh Luran Plain, Khuzistan (Helbaek 1969). Sabz phase and Khazineh phase (5500–5000 bc). Scarce remains: brittle einkorn wheat (few); einkorn wheat (few); emmer wheat (few); free-threshing wheat (frequent); two-rowed barley (prevailing); six-rowed barley (frequent); naked barley (few); lentil (frequent); grass pea (few); flax (frequent). Wild: *Triticum boeoticum* (few); *Avena* (few); *Capparis* (few); *Prosopis* (frequent); *Amygdalus* (few); *Pistacia* (few); legumes (frequent); other wild grasses (few).

3. **Tepe Musiyan**, Deh Luran Plain, Khuzistan (Helbaek 1969). Mehmeh phase (4500–4000 bc). Scarce remains: einkorn wheat (few); emmer wheat (frequent); free-threshing wheat (rare); six-rowed barley (frequent); naked barley (rare); lentil (few); flax (few). Wild: *Avena* (rare); *Lolium* (few).

4. **Tepe Hasanlu**, Solduz Valley (Tosi 1975). (i) Hajji Firuz Tepe, periods X–VIII (6th–4th millennia bc). Unspecified quantities: emmer wheat (frequent); two-rowed barley (prevailing). (ii) Pisdeli Tepe, period VIII (4th millennium bc). Rich remains: emmer wheat (frequent); free-threshing wheat (frequent); two-and six-rowed barley (prevailing).

5. **Tepe Yahya** and adjacent sites, Dowlatabad Plain 200 km south of Kerman (Costantini and Costantini-Biasini 1985). Neolithic, periods VII and VI (late 6th millennium bc to 5th millennium bc). Rich remains, unspecified quantities. The prevailing crops are einkorn wheat, emmer wheat, two-rowed barley, and six-rowed barley. Free-threshing wheat and a rounded, small-grained 'sphaerococcum' like barley are also present.

Iraq

(General references: Braidwood 1960; Renfrew 1984; Miller 1991.)

1. **Jarmo**, Kurdistan (Helbaek 1959b, 1960a, 1966d; Braidwood 1960). (*c* 6750 bc). Scarce remains: (both imprints and carbonized remains): brittle and non-brittle einkorn wheat (rare); brittle and non-brittle emmer wheat (frequent); brittle two-rowed barley (frequent); non-brittle two-rowed barley (few); lentil (rare); pea (rare). Wild: *Pistacia*: *Prosopis*; *Aegilops*; *Lathyrus*.

2. **Tell es-Sawwan**, Samarra (Helbaek 1964b). (1st half of 6th millennium bc). Numerous remains: einkorn wheat (few); emmer wheat (frequent); free-threshing wheat (few); two-rowed and six-rowed barley (prevailing); naked barley (frequent); flax (rare). Wild: *Prosopis farcta* (frequent); *Capparis spinosa* (frequent).

3. **Yarym Tepe**, northern Iraq (Bakhteyev and Yanushevich 1980).

(i) Yarym Tepe I, Neolithic (6th millennium bc). Numerous remains: emmer wheat (few); free-threshing wheat (few); spelta wheat? (rare); hulled two-rowed barley (rare); hulled six-rowed barley (prevailing); naked barley (few). (ii) Yarym Tepe II Neolithic (5th millennium bc). Rich remains: emmer wheat (few); free-threshing wheat (rare); spelta wheat? (few); hulled six-rowed barley (prevailing); naked barley (few). No other plants mentioned. (Note that the identification of spelta wheat is based on kernel morphology only. It cannot be regarded as definite.)

4. **Choga Mami**, Mandali (Helbaek 1972). (i) Samarra phase (second half of 6th millennium bc). Rich remains: brittle einkorn wheat (rare); einkorn wheat (few); emmer wheat (frequent); free-threshing wheat (frequent); brittle two-rowed barley (frequent); two-rowed and six-rowed barley (rare); naked barley (frequent); lentil (frequent); pea (few); flax (frequent). Wild: *Avena* sp. (few); *Lolium* sp. (prevailing); other grasses (frequent); *Pistacia atlantica* (few). (ii) Post-Samarra phase (*c.* 5000 bc). Numerous remains: brittle einkorn and einkorn wheat (rare); emmer wheat (few); free-threshing wheat (rare); six-rowed barley (few); naked barley (rare); lentil (few); flax (rare). Wild: *Lolium* sp. (prevailing); *Pistacia atlantica* (few).

5. **Shahrzoor Valley** near Sulaymaniya (Helbaek 1960b). (i) Tell Chragh (early 4th millennium bc). Scarce remains: emmer wheat (rare); two-rowed barley (prevailing); lentil (rare). Wild: *Avena* sp. (few); *Lolium* sp. (frequent); *Saponaria vaccaria* (few). (ii) Tell Qurtass (end of 3rd millennium bc). Numerous remains: emmer wheat (few); free-threshing wheat (rare); lentil (prevailing); bitter vetch (frequent).

Turkey

(General reference: Miller 1991.)

1. **Çayönü** near Diyarbakir (van Zeist 1972). Phases 1–3 contain material which still retains wild-type morphology. (i) Phase 1 (7500 bc). Scarce remains: brittle einkorn and einkorn wheats (few); brittle emmer wheat (few); emmer wheat (frequent); brittle two-rowed barley (rare); bitter vetch (prevailing); pea (frequent); chickpea (rare). Wild: *Vicia* sp. (frequent); *Pistacia atlantica* (frequent); *Amygdalus* sp. (few); *Lolium* sp. (few). (ii) Phase 2 (*c.* 7200 bc). Scarce remains: brittle einkorn and einkorn wheats (rare); brittle emmer and emmer wheats (rare); brittle barley (rare); lentil (few); pea (frequent); bitter vetch (prevailing). Wild: *Lathyrus* sp. (rare); *Vicia* sp. (frequent); *Linum* sp. (frequent); *Pistacia atlantica* (frequent) *Amygdalus* sp. (rare); *Lolium* sp. (rare). (iii) Phase 3 (*c.* 7000 bc). Scarce remains: brittle einkorn and einkorn wheats (rare); emmer wheat (prevailing); lentil (few); pea (few); bitter vetch (prevailing). Wild: *Vicia* sp. (frequent); *Pistacia atlantica* (frequent); *Amygdalus* sp.

(frequent); *Quercus* acorns (rare); wild rye grass (rare). (iv) Phase 4 (*c.* 6800 bc). Numerous remains: cultivated lentil (prevailing); cultivated pea (frequent); bitter vetch (frequent); chickpea (few). Wild: *Pistacia atlantica* and *Celtis tournefortii* (rare). (v) Phase 5 (*c.* 6500 bc). Numerous remains: cultivated lentil (prevailing); cultivated pea (frequent); bitter vetch (few). Wild: *Pistacia atlantica* (rare).

2. **Can Hasan**, Konya Plain (Renfrew 1968; French *et al.* 1972; Hillman 1978). (i) Can Hasan III, Aceramic Neolithic (middle of 7th millennium bc). Numerous remains: brittle einkorn wheat (rare); einkorn wheat (few); emmer wheat (frequent); free-threshing wheat (prevailing); two-rowed barley (frequent); naked barley (rare); rye (few); lentil (few); bitter vetch (frequent); other vetches (frequent). Wild: *Juglans regia* (rare); *Celtis tournefortii* (frequent); *Vitis* (rare); *Prunus* sp. and *Crataegus* sp. (rare); *Lithospermum* sp. (prevailing). (ii) Can Hasan I, Konya Plain (*c.* 5250 bc). Scarce remains: wheat (few); six-rowed barley (frequent); pea (frequent).

3. **Hacilar**, Konya Plain (Helbaek 1970). (i) Aceramic Neolithic (*c.* 6750 bc). Scarce remains: brittle einkorn wheat (frequent); emmer wheat (prevailing); brittle barley (frequent); naked barley (few); lentil (rare). (ii) Late Neolithic (*c.* 5400–5050 bc). Rich remains: brittle einkorn wheat (frequent); einkorn wheat (frequent); emmer wheat (few); free-threshing wheat (few); two-rowed barley (rare); six-rowed barley (few); naked barley (prevailing); lentil (few); bitter vetch (few). Wild: *Pisum* sp. (frequent); *Aegilops umbellulata* (rare); *Pistacia atlantica* (rare); *Celtis australis* (frequent); *Capparis spinosa* (few); *Malus* sp. (rare); *Amygdalus* sp. (rare).

4. **Çatal Hüyük**, Konya Plain (Helbaek 1964a) 5850–5600 bc. Rich remains: einkorn wheat (frequent); emmer wheat (prevailing); free-threshing wheat (few); naked barley (frequent); pea (frequent); bitter vetch (few). Wild: *Quercus* acorns (frequent); *Pistacia atlantica* (rare); *Celtis australis* (frequent); *Amygdalus* sp. (rare); *Pisum elatius* (rare); *Vicia* sp. (frequent).

5. **Erbaba**, Beysehir, south-central Anatolia (van Zeist and Buitenhuis 1983). Neolithic (5800–5400 bc). Rich remains: einkorn wheat (frequent); emmer wheat and free-threshing wheat (prevailing); spelta wheat? (rare); hulled two-rowed barley (frequent); naked barley (frequent); lentil (frequent); pea (co-prevailing); bitter vetch (frequent). Wild: *Lathyrus* cf. *cicera*; *Triticum boeoticum* and several herbaceous plants and weeds.

6. **Korucutepe**, Altinova Plain (van Zeist and Bakker-Heeres 1975b). (i) Chalcolithic (4500–4000 bc). Scarce remains: emmer wheat (prevailing); free-threshing wheat (few); two-rowed barley (few); lentil (rare); flax (rare). Wild: *Vitis* (few); *Capparis* (rare); *Bromus* sp. (few); *Lolium* and *Onopordon* (rare). (ii) 4000–3500 bc. Scarce remains: einkorn wheat (rare); emmer wheat (prevailing); free-threshing wheat (few); two-rowed barley (few); naked barley (few); bitter vetch (few); flax (few). Wild: *Vitis*

(few); *Lolium* sp. (few). (iii) 3500–3000 bc. Scarce remains: emmer wheat (few); two-rowed barley (frequent); chickpea (rare); flax (prevailing). Wild: *Crataegus azarolus* (few); *Lolium* sp. (few). (iv) 2600–2300 bc. Rich remains: emmer wheat (few); free-threshing wheat (frequent); two-rowed barley (prevailing); naked barley (rare); lentil (few); chickpea (frequent). Wild: *Pistacia atlantica* (rare); *Vitis* (rare).

   7. **Girikihaciyan**, near Diyarbakir (van Zeist 1979–80). Halafian (5000– 4500 bc). Numerous remains: einkorn wheat (rare); emmer wheat (prevailing); free-threshing wheat (rare); two-rowed barley (few); lentil (frequent); chickpea (few); bitter vetch (frequent); flax (few). Wild: *Pistacia atlantica* (rare); *Amygdalus* sp. and *Crataegus* sp. (rare); *Vicia* sp. (few); *Lolium* sp. (few).

Cyprus

(General refrences: Kroll 1991; Miller 1991; Hansen 1991.)

   1. **Cape Andreas-Kastros**, Aceramic Neolithic (van Zeist 1981) (*c.*6000 bc). Numerous remains: einkorn wheat (frequent); emmer wheat (prevailing); six-rowed barley (frequent); lentil (frequent); pea? (rare); flax (rare). Wild: *Vicia faba/narbonensis* (rare); *Olea europaea* (rare); *Pistacia atlantica/ terebinthus* (rare); *Ficus carica* (rare); *Lolium* sp. (frequent); *Vicia* sp. (few); *Malva* sp. (few).

   2. **Khirokitia** (Waines and Stanley Price 1975–77; Miller 1984; Hansen 1991), Aceramic Neolithic (*c.* 5500 bc). Numerous remains: einkorn wheat (prevailing); emmer wheat (frequent); six-rowed barley (frequent); lentil (frequent); pea (few). Wild: *Lathyrus* cf. *sativus* (rare); *Vicia faba/narbonensis* (rare); *Ficus carica* (frequent); *Olea europaea* (rare); *Prunus insititia* (few); *Pistacia* sp. (rare).

   3. **Dhali Agridhi**, Idalion (Stewart 1974) (*c.* 4500 bc; or early 4th millennium bc). Numerous remains: emmer wheat (few); free-threshing wheat (rare); two and six-rowed barley (frequent); naked barley (few); lentil (frequent); pea (rare). Wild: *Olea europaea* (frequent); *Vitis vinifera* (few); *Ficus carica* (frequent); *Amygdalus* sp. (rare); *Pistacia* sp. (rare); *Anchusa* sp. (frequent).

   4. **Hala Sultan Tekke**, near Larnaca (Hjelmqvist 1979b), Bronze Age (*c* 1200 bc). Rich remains: emmer wheat (few imprints); free-threshing wheat (rare); hulled and naked barley (prevailing – imprints); lentil (rare); olive (few); grape vine pips (frequent); fig pips (frequent); pomegranate (rare); citron (rare). Wild: *Pistacia terebinthus* (rare); *Capparis spinosa* (rare); *Zizyphus spina-christi* (rare); *Lupinus albus* (rare); *Chrozophora verbascifolia* (few); *Onopordon* cf. *illyricum* (few).

Syria

(General reference; Miller 1991.)

1. **Tell Abu Hureyra**, Northern Syria (Hillman 1975). (i) Epipalaeolithic (9200–8500 bc). Rich remains. Wild: one- and two-grained brittle einkorn wheat (prevailing); brittle barley (few); brittle rye (few); wild lentil (few); *Pistacia* sp. (frequent); *Celtis* (frequent); *Capparis spinosa* (rare); numerous grass and herb seeds. (ii) Aceramic Neolithic (Pre-Pottery Neolithic B) (7500–6500 bc). Rich finds: two-grained brittle einkorn wheat (rare); einkorn wheat (few); brittle emmer wheat (rare); emmer wheat (frequent); brittle two-rowed barley (few); six-rowed barley (few); naked barley (frequent); lentil (frequent); chickpea (few); *Vicia faba*-type (rare). Wild: *Avena* (rare); *Secale* (rare); *Vitis vinifera* (rare); *Capparis spinosa* (frequent); *Prosopis farcta* (abundant); *Lithospermum* and *Echium* (abundant); numerous grass and herb seeds.

2. **Tell Mureybit**, Northern Syria (van Zeist and Bakker-Heeres 1986b). Final Mesolithic (8050–7550 bc). Rich remains. Wild: brittle einkorn wheat (prevailing); brittle barley (few); wild lentil (few); wild bitter vetch (rare); wild flax (rare); *Pistacia atlantica* (rare); numerous grass and herb seeds.

3. **Tell Aswad**, 30 km south-east of Damascus (van Zeist and Bakker-Heeres 1985). (i) Phase IA (7800–7600 bc). Numerous remains: emmer wheat (frequent); two-rowed or brittle two-rowed barley (few); lentil (rare); pea (few). Wild: *Pistacia* sp. (prevailing); *Ficus carica* (frequent); *Capparis spinosa* (rare); *Amygdalus* sp. (rare); *Trigonella* sp. (frequent); numerous grass and herb seeds. (ii) Phase IB (7600–7300 bc). Numerous remains: emmer wheat (frequent); two-rowed or brittle two-rowed barley (rare); lentil (rare); pea (rare). Wild: *Pistacia* sp. (frequent); *Trigonella* sp. (prevailing); *Ficus carica* (few); *Capparis spinosa* (rare); numerous herb and grass seeds. (iii) Phase II East (6925–6600 bc). Rich remains: einkorn wheat (rare); emmer wheat (prevailing); free-threshing wheat (few); two-rowed barley (few); naked barley (rare); lentil (rare); pea (few). Wild: *Pistacia* sp. (frequent); *Trigonella* sp. (frequent); *Ficus carica* (rare); numerous grass and herb seeds; (iv) Phase II West (6925–6600 bc). Rich remains: einkorn wheat (rare); emmer wheat (prevailing); free-threshing wheat (few); two-rowed barley (frequent); naked barley (rare); pea (few); lentil (few); flax (rare). Wild: *Pistacia* sp. (frequent); *Ficus carica* (frequent); *Capparis spinosa* (rare); *Vitis vinifera* (rare); *Trigonella* sp. (frequent); *Cyperus* sp., *Carex* sp., *Lolium* sp. and *Phalaris* sp. (frequent); numerous other grass and herb seeds.

4. **Tell Ramad** 20 km south-west of Damascus (van Zeist and Bakker-Heeres 1985). Pre-Pottery Neolithic B. (i) Phase I (*c.* 6200 bc). Numerous remains: einkorn wheat (few); emmer wheat (most frequent); free-threshing wheat (rare); two-rowed barley (rare); lentil (frequent); pea (few). (ii) Phase

II (*c*. 5950 bc). Rich remains: einkorn wheat (rare); emmer wheat (frequent); two-rowed barley (few); naked barley (rare); lentil (few); pea (few); flax (few). Wild: *Pistacia atlantica* and numerous grass and leguminous seeds.

5. **Tell Bouqras**, edge of the Euphrates Valley, eastern Syria. (van Zeist and Waterbolk-van Rooijen 1985). Pre-Pottery Neolithic B (6300–5800 bc). Numerous remains: einkorn wheat (few); emmer wheat (few); free-threshing wheat (common); hulled barley (rare); naked barley (common); lentil (few); pea (rare). Remains of numerous species of herbs, shrubs and trees including seeds of *Ficus carica*, *Pistacia* and wild *Linum*.

6. **Ras Shamra**, north of Latakia (van Zeist and Bakker-Heeres 1986a). Neolithic level (6500–5250 bc). Rich remains: einkorn wheat (few); emmer wheat (prevailing); free-threshing wheat (rare); hulled two-rowed barley (frequent); lentil (frequent); pea (rare); grass pea (rare); flax (rare). Numerous seeds of wild plants.

Israel and Jordan

(General reference: Miller 1991)

1. **Jericho**, Lower Jordan Valley (Hopf 1983). (i) Pre-Pottery Neolithic A (*c*. 8000–7300 bc). Scarce remains: emmer wheat (rare); two-rowed barley (prevailing). Wild: *Lens* sp. (few); few fragments of *Pistacia* and *Amygdalus*. (ii) Pre-Pottery Neolithic B (*c*. 7300–6500 bc). Rich remains: einkorn wheat and emmer wheat (prevailing); two-rowed barley (frequent); lentil (frequent); pea (frequent); *Vicia faba*-type (few); chickpea (rare); flax (an imprint of a single capsule). (iii) Ceramic Neolithic (first half of 5th millennium bc). Rich remains: einkorn wheat and emmer wheat (frequent); two-rowed barley (prevailing). Wild: some grass and herb seeds. (iv) Chalcolithic (early 4th millennium bc). Numerous remains: wheat (rare); two-rowed barley (prevailing). Wild: *Ficus carica* (rare). (v) Early Bronze Age (*c*. 3200 bc). Rich remains: einkorn wheat (frequent); emmer wheat (prevailing); free-threshing wheat (frequent); two-rowed and six-rowed barley (frequent); lentil (frequent); chickpea (few); flax, a capsule and seed (rare); grape vine (frequent, both pips and berries); date palm (rare); fig (rare). Wild: *Allium* cf. *ampeloprasum* (few); *Pistacia atlantica* (rare); numerous grass and herb seeds. (vi) Middle Bronze Age. Rich remains: emmer wheat (frequent); free-threshing wheat (frequent); six-rowed barley (prevailing); lentil (rare); pea (rare); *Vicia faba*-type (rare); grape vine (rare); fig (rare); pomegranate (rare). Wild: *Allium* cf. *Ampeloprasum* (rare); some grass and herb seeds.

2. **Ain Ghazal**, north-east of Amman (Rollefson and Simmons 1985). Pre-Pottery Neolithic B (7200–6000 bc). Rich remains: emmer wheat (frequent); two-rowed barley (frequent); lentil (frequent); pea (frequent); chickpea (few); flax (rare).

3. **Tuleilat Ghassul**, Lower Jordan Valley (Zohary and Spiegel-Roy 1975). Chalcolithic (*c*. 3700 bc). Rich remains: emmer wheat (few); two and six-rowed barley (frequent); lentil (few); olive (frequent); date palm (few).

4. **Arad**, Northern Negev (Hopf 1978a). Early Bronze Age (*c*. 2770 bc). Rich remains: einkorn wheat (rare); emmer wheat (frequent); free-threshing wheat (rare); two-rowed barley (prevailing); six-rowed barley (frequent); lentil (frequent); pea (few); chickpea (few); bitter vetch (rare); broad bean (rare); linseed (few); olive (frequent); grape vine (frequent); pomegranate (rare). Wild: *Pistacia atlantica* (rare); *Quercus* acorns (rare); *Lolium temulentum* as well as other grass and herb seeds.

5. **Bab edh-Dhra**, south-east of Dead Sea (McCreery 1979). Early Bronze Age. Rich remains: emmer wheat (probably also einkorn and free-threshing wheat) (few); two-rowed barley (few); six-rowed barley (prevailing); lentil (few); chickpea (frequent); linseed (few); olive (frequent); grape vine (frequent, whole berries and pips); fig (frequent); almond (rare). Wild: *Pistacia atlantica* (few); *Prunus insititia*-like (rare); *Lolium temulentum* and herb seeds. Textiles made of flax fibres.

Egypt

(General references: Täckholm 1976; Germer 1985.)

1. **Merimde**, Beni Salâme, western Nile delta. (i) Early Neolithic, late 6th and 5th millennia bc (M. Hopf, unpublished data). Rich remains: emmer wheat (prevailing); free-threshing wheat? (rare); hulled six-rowed barley (frequent); lentil (few) pea (few); flax (rare). Wild: *Lolium* and several other weeds, sedges, and legumes (frequent). (ii) Middle Neolithic (Helbaek 1955). Rich remains: emmer wheat (prevailing); free-threshing wheat (rare and later absent); hulled six-rowed barley (frequent).

2. **Fayum**, (Caton-Thompson and Gardner 1934, pp 46–9; Stemler 1980). Neolithic (5th millennium bc). Large quantities of parched and of charred grains in underground silos: mainly emmer wheat and two-rowed and six-rowed barley; also flax textile.

3. **Nagada-Khattara** region Upper Egypt (Wetterstrom, personal communication). (i) Site KH3 (*c*. 3800 bc). Numerous remains: emmer wheat (prevailing); hulled six-rowed barley (frequent); pea (rare); flax (rare). Wild: *Citrullus colocynthis* (rare); *Zizyphus spina-christi* (rare). (ii) South Town (*c*. 3400 bc). Numerous remains: emmer wheat (prevailing); hulled six-rowed barley (frequent); pea (rare); bitter vetch (rare); flax (few). Wild: *Citrullus colocynthis* (few); *Zizyphus spina-christi* (rare).

4. **Kom el-Hisn**, west part of the Nile delta (Moens and Wetterstrom 1988). Old Kingdom deposits. Numerous remains: emmer wheat (prevailing); barley (frequent); pea (rare); flax (rare); clover seeds, probably

Trifolium alexandrinum (co-prevailing). Wild: *Lolium temulentum* and *Phalaris paradoxa* (frequent); numerous herbs, reeds and sedges.

5. **Saqqara**, Memphis. (i) Djoser pyramid, 3rd dynasty, 2630 bc. (Lauer *et al.* 1950). Rich remains (desiccated material); emmer wheat and hulled six-rowed barley (the bulk of the find); lentil (rare); sycamore fig (numerous dry sycons). Apparently imported: fragments of grape vine raisins (few); *Mimusops schimperi* (rare); *Juniperus oxycedrus* (few). Wild: *Lolium temulentum* (the commonest weed); *Phalaris paradoxa* (rare); *Vicia sativa* (few); *V. lutea* (few); *V. narbonensis* (rare); *Lathyrus aphaca* (few); *L. marmoratus* (rare); *L. hirsutus* (few); *Scorpiurus muricata* (few); *Trigonella hamosa* (rare); *Medicago hispida* (rare); *Rumex dentatus* (few); *Anthemis pseudocotula* (rare); *Beta vulgaris* (rare); *Zizyphus spina-christi* (numerous); *Acacia nilotica* (rare); *Balanites aegyptiaca* (rare). (ii) Queen Icheti's tomb, 6th dynasty, 2550 bc (Helbaek 1953). Rich remains (mummified material): emmer wheat prevailing.

6. **Tutankhamun tomb**, Valley of the Kings, near Thebes, (Germer 1989a; Hepper 1990) *c.*1325 BC. A very rich find of desiccated, excellently preserved plants and plant products: emmer wheat (main element in the model granary); six-rowed barley (large quantities, including germinated seeds in 'Osiris bed'); lentil (few); chickpea (few); fenugreek (in several baskets, mixed with coriander); flax (linen, arrow strings, few seeds); olives (leaves in garlands, jars of olive oil); grape wine (shrivelled berries and their pips, wine amphoras with records on production sites and dates); sycamore fig (fruits and timber); date palm (fruits, ropes); almond (basket with fruits); watermelon (numerous seeds); garlic (numerous bulbs, some with leaves); black cumin (numerous seeds); coriander seeds (main item in several baskets); safflower (few seeds, textiles dyed with safflower pigment). Remains of several ornamental plants and plants collected from the wild. Timber (in furniture and implements) of numerous, both local and foreign, trees, including imported cedar (*Cedrus libani*) and ebony (*Dalbergia melanoxylon*).

Caucasia, Transcaucasia, and central Asia

(General reference: Wasylikowa *et al.* 1991.)

1. **Chokh**, Dagestan (Lisitsina 1984; Schultze-Motel 1989). Neolithic (6th millennium bc) Unspecified quantities: einkorn wheat; emmer wheat; free-threshing wheat; hulled barley; naked barley; several wild grasses and trees.

2. **Arukhlo 1 and Arukhlo 2**, Bolnisskiy region, Georgia (Janushevich 1984; Lisitsina 1984; Schultze-Motel 1988). Eneolithic (5th millennium bc). Unspecified quantities (except for wheats), mostly imprints: einkorn wheat (few); emmer wheat (numerous); free-threshing wheat (prevailing);

spelta wheat (rare); two-rowed and six-rowed hulled barley; naked barley; broomcorn millet; lentil; pea; bitter vetch.

3. **Imiris-Gora**, Marnsulskij region, Georgia (Lisitsina and Prishche-penko 1977; Schultze-Motel 1988). Eneolithic (5th–4th millennia bc). Imprints, unspecified quantities: emmer wheat; free-threshing wheat; spelta wheat?; six-rowed hulled and naked barley; broomcorn millet?; *Avena* sp.

4. **Djeitun** and several other sites in south Turkmenia including Altyn Tepe (Masson and Sarianidi 1972; Janushevich 1984; Hillman and Charles, in press). Djeitun culture (5050 bc). Unspecified quantities: einkorn wheat (prevailing); emmer wheat (few); free-threshing wheat? (rare); hulled and naked barley (co-prevailing). Wild: numerous herbaceous and annual plant species.

India and Pakistan

(General References: Costantini and Costantini-Biasini 1985; Kajale 1991; Weber 1991.)

1. **Mehrgarh**, west margin of the Indus plain, Baluchistan (Jarrige and Meadow 1980; Costantini 1984). (i) Period I, Neolithic (6th millennium bc). Scarce remains (imprints, unspecified quantities): einkorn wheat; emmer wheat; free-threshing wheat, two-rowed and six-rowed barleys; naked barley. Wild: *Zizyphus jujuba* and *Phoenix dactylifera*. (ii) Period II, Neolithic (5th millennium bc). Numerous remains (unspecified quantities): wheats; barleys; cotton seeds (either wild or cultivated). (iii) Period III (*c.* 4000 bc). Numerous remains (unspecified proportions): einkorn wheat; emmer wheat; free-threshing wheat including Indian wheat *Triticum aestivum* subsp. *sphaerococcum*; two-rowed and six-rowed barleys; naked barley; *Avena* (probably wild).

2. **Pirak**, Baluchistan (Costantini 1979). Period I (*c.* 1600 bc). Rich remains (unspecified quantities): free-threshing wheat including Indian wheat; two-rowed barley; six-rowed barley; naked barley; rice, broomcorn millet; sorghum; chickpea; flax; grape vine. Wild: *Citrullus* cf. *colocynthis, Zizyphus jujuba*.

3. **Harappa** and several other sites of the Harappan culture in Pakistan and north-west India (Kajale 1991; Vishnu-Mittre and Savithri 1982); Harappan (2250–1750 bc). Rich remains (unspecified quantities): free-threshing wheat (including Indian wheat *Triticum aestivum* subsp. *sphaero-coccum*); hulled and naked six-rowed barley; rice; pea; lentil; chickpea; sesame; cotton; several south Asian pulses ('grams'): *Vigna radiata, V. mungo*; three African millets (*Sorghum bicolor, Pennisetum typhoides* and *Elusine coracana*).

Greece

(General references: Renfrew 1979; Kroll 1991)

1. **Franchthi Cave**, Argolis (Hansen 1978, 1992). (i) Palaeolithic (12 000 –7000 bc). Scarce remains: Wild: brittle two-rowed barley (few); *Lens* sp. (few) *Avena* sp. (rare); *Vicia* sp. (few); *Pistacia* sp. (few); *Amygdalus* sp. (few); *Lithospermum, Alkanna, Anchusa* (frequent). (ii) Mesolithic (7300–6000 bc). Scarce remains: Wild: brittle two-rowed barley (few); *Lens* sp. (few); *Pisum* sp. (few); *Avena* sp. (few); *Pistacia* sp., *Amygdalus* sp. *Pyrus* sp. (few); *Lithospermum, Alkanna* and *Anchusa* (frequent). (iii) Early Neolithic (6000–5300 bc). Scarce remains: emmer wheat (few); two-rowed barley (few). Wild: *Pistacia* sp. (few); *Amygdalus* sp. (rare); *Lithospermum, Alkanna, Anchusa, Lens* sp., *Vicia* sp. (few). (iv) Middle Neolithic (5000–4300 bc). Scarce remains: einkorn wheat (rare); emmer wheat (frequent); two-rowed barley (frequent); lentil (few); bitter vetch (rare). Wild: *Pistacia* sp. and *Amygdalus* sp. and *Pyrus* sp. (rare); *Lithospermum, Alkanna, Anchusa* (rare). (v) Late Neolithic (4300–2800 bc). Scarce remains: einkorn wheat (frequent); emmer wheat (frequent); two-rowed barley (frequent); lentil (few); bitter vetch (rare). Wild: *Pistacia* sp., *Pyrus* sp., *Amygdalus* sp., *Vitis sylvestris* (rare); *Lithospermum, Alkanna* and *Anchusa* (rare).

2. **Sesklo** (including Argissa, and Otzaki), Thessaly (Kroll 1981a; Hopf 1962). (i) Aceramic and Early Neolithic (end of 7th and beginning of 6th millennium bc). Numerous remains: einkorn wheat (frequent); emmer wheat (prevailing); six-rowed barley (frequent); lentil (frequent); bitter vetch (few). Wild: *Ficus carica* (frequent); *Vitis vinifera* (rare); *Sambucus ebulus* (rare); *Pistacia atlantica* (frequent); *Lithospermum arvense* (frequent); *Avena* sp. (frequent). (ii) Proto-and pre-Sesklo phases (6th millennium bc). Rich remains: einkorn and emmer wheat (prevailing); free-threshing wheat (rare); two-rowed barley (few); six-rowed hulled and naked barley (frequent); lentil (frequent); pea (few); bitter vetch (few); chickpea (rare); flax (few). Wild: *Ficus carica* (frequent); *Sambucus ebulus* (rare); *Pistacia atlantica* (rare); *Lithospermum arvense* (frequent); *Avena* sp. (few). (iii) Sesklo phase (5th millennium bc). Numerous remains: einkorn wheat (frequent); emmer wheat (prevailing); six-rowed barley (few); broomcorn millet (rare); lentil (frequent); pea (rare); bitter vetch (rare); flax (rare). Wild: *Ficus carica* (frequent); *Vitis vinifera* (rare); *Pistacia atlantica* (rare); *Lithospermum arvense* (frequent); *Avena* sp. (few); (iv) Dimini phase, Late Neolithic (early to middle of 4th millennium bc). Numerous remains: einkorn wheat (frequent); emmer wheat (prevailing); free-threshing wheat (rare); six-rowed barley (frequent); lentil (few); faba bean; bitter vetch (rare); flax (rare). Wild: *Ficus carica* (frequent); *Vitis vinifera* (frequent); *Amygdalus* sp. (rare); *Lithospermum arvense*

(rare); *Avena* sp. (few). (v) Rachmani phase, Late Neolithic (late 4th to early 3rd millennium bc). Rich remains: einkorn wheat, emmer wheat, six-rowed barley and lentil (equally frequent); free-threshing wheat (rare); naked barley (rare); pea (rare); grass pea (few); flax (rare). Wild: *Ficus carica* (frequent); *Vitis vinifera* (rare); *Quercus* acorns (rare); *Amygdalus* sp. (rare); *Camelina sativa* (a single seed); *Lithospermum arvense* (few); *Avena* sp. (few).

3. **Nea Nikomedeia**, Macedonia (van Zeist and Bottema 1971). Early Neolithic (*c.* 5470 bc). Rich remains: einkorn wheat (frequent); emmer wheat (prevailing); naked barley (frequent); lentil (frequent); pea (frequent); bitter vetch (frequent). Wild: *Quercus* acorns (few); *Cornus mas* (rare); *Prunus* cf. *spinosa* (rare).

4. **Knossos**, Crete (Renfrew 1979, table 2). Early Neolithic (*c.* 6000 bc). Numerous remains: emmer wheat (few); free-threshing wheat (prevailing); barley (few).

5. **Dimini**, Thessaly (Kroll 1979). Late Neolithic (Classic phase). Rich remains: einkorn wheat (rare); emmer wheat (frequent); six-rowed barley (rare); naked barley (frequent); lentil (frequent); pea (frequent); faba bean (few); bitter vetch (few); chickpea (few); grass pea (frequent). Wild: *Vitis vinifera* (rare); *Amygdalus* sp. (few).

6. **Lerna**, Argolis (Hopf 1961b). (i) Final Neolithic: *Ficus carica, Arbutus*. (ii) Early Bronze Age. (3rd millennium bc). Numerous remains: einkorn and emmer wheats (few); six-rowed barley and naked barley (few); lentil (frequent); pea (few); broad bean (frequent); bitter vetch (frequent); grass pea? (few); linseed (frequent); fig (frequent, both fruits and pips); grape vine (few).

7. **Kastanas**, Macedonia (Kroll 1983, 1984). (i) Early Bronze Age (second half of 3rd millennium bc). Rich remains: einkorn wheat (frequent); emmer wheat (frequent); free-threshing wheat (rare); spelta wheat (rare); six-rowed barley (prevailing); lentil (frequent); pea (rare); bitter vetch (frequent); faba bean (rare); grass pea (rare); linseed (few); poppy (rare); grape vine (few); fig (frequent). Wild: *Quercus* acorns (few); *Pyrus* sp. (few); *Rubus fruticosus* (rare); *Sambucus ebulus* (few); *Cornus mas* (few); *Lolium temulentum* (few); other grass and herb seeds. (ii) Late Bronze Age (second half of 2nd millennium bc). Rich remains: einkorn and emmer wheats (frequent); spelta wheat (rare); free-threshing wheat (rare); six-rowed barley (few); lentil (frequent); bitter vetch (prevailing); broomcorn millet (frequent); foxtail millet (few); poppy (rare); grape vine (frequent); fig (few). Wild: *Quercus* acorns (rare); *Sambucus ebulus* (rare); *Secale cereale and Avena* sp. (contaminating wheat and barley); *Camelina sativa* (rare); *Fragaria* sp. (rare); numerous grass and herb seeds.

Yugoslavia

(General references: Renfrew 1979; Kroll 1991.)

1. **Starčevo**, Serbia (Renfrew 1979, table 5). Early Neolithic (*c*. 5000 bc). Numerous remains: einkorn wheat (few); emmer wheat (prevailing); naked wheat (few); six-rowed barley (few); pea (few). Wild: *Malus* sp. and *Cornus mas* (few).

2. **Anza**, Macedonia (Renfrew 1976). (i) Early Neolithic (5300–4750 bc). Numerous finds: einkorn wheat (few); emmer wheat (frequent); free-threshing wheat (few); six-rowed barley (few); lentil (rare); pea (rare). Wild: *Malus* sp., *Corylus avellana, Cornus mas* and *Vitis vinifera* (few). (ii) Late Neolithic (4100 bc). Scarce remains: einkorn and emmer wheats (frequent); six-rowed barley (rare); lentil (rare); pea (rare).

3. **Obre**, Bosnia-Hercegovina (Renfrew 1974). (i) Early Neolithic (Starčevo group, 4760 bc). Scarce remains: einkorn wheat (rare); emmer wheat (prevailing); free-threshing wheat (rare); lentil (few); pea (frequent). Wild: *Cornus mas* (rare). (ii) Late Neolithic (Butmir culture, 4000–3860 bc). Numerous remains: einkorn wheat (frequent); emmer wheat (frequent); free-threshing wheat (rare); six-rowed barley (frequent); naked barley (frequent); lentil (frequent).

4. **Vršnik**, Štip, Macedonia (Hopf 1961a). Early Neolithic (Starčevo group, 4900 bc). Numerous remains: einkorn wheat (prevailing); emmer wheat (frequent); free-threshing wheat (rare); barley (rare).

5. **Pokrovnik**, Dalmatia (Karg and Müller 1990). Middle Neolithic, Danilo culture (*c*. 4350 bc). Numerous remains in a storage jar: einkorn wheat (few); emmer wheat (prevailing).

6. **Gomolava**, Serbia (van Zeist 1975). Late Neolithic (Vinča culture, 3800–3500 bc). Rich remains: einkorn wheat (prevailing); emmer wheat (frequent); six-rowed barley (frequent); broomcorn millet (frequent); lentil (few); pea (rare); bitter vetch (rare); flax (rare). Wild: *Malus* sp., *Cornus mas, Vitis vinifera, Fragaria vesca* (few); *Avena* sp. (rare).

Bulgaria

(General references: Renfrew 1979; Kroll 1991.)

1. **Karanovo Mogila**, Sliven (Hopf 1973b; Renfrew 1979, table 7). (i) Early Neolithic Phase I (4800 bc). Scarce remains: einkorn wheat (frequent); emmer wheat (prevailing). (ii) Early Neolithic Phase II (4623 bc). Scarce remains: einkorn wheat (frequent); emmer wheat (prevailing), (iii) Middle Neolithic Phase III (4410 bc). Scarce remains: einkorn wheat (few); emmer wheat (prevailing); naked six-rowed barley (rare). (iv) Eneolithic Phase V. Rich remains: einkorn wheat (few); emmer wheat

(few); naked six-rowed barley (prevailing); lentil (rare); bitter vetch (rare). (v) Eneolithic Phase VI (3890 bc). Rich remains: einkorn wheat (frequent); emmer wheat (frequent); naked six-rowed barley (few); bitter vetch (prevailing).

2. **Azmaška Mogila**, Sliven (Hopf 1973b; Renfrew 1979, table 7). (i) Neolithic Phase I (4928–4770 bc). Rich remains: einkorn wheat (few); emmer wheat (prevailing); free-threshing wheat (frequent); naked six-rowed barley (rare); lentil (frequent); pea (few); grass pea (few). (ii) Middle Neolithic Phase II. Rich remains: bitter vetch (pure). Wild: *Sambucus* sp. (iii) Eneolithic (*c.* 3640 bc). Rich remains: einkorn wheat (frequent); emmer wheat (frequent); naked six-rowed barley (frequent); hulled six-rowed barley (rare); pea (few); lentil (frequent); bitter vetch (prevailing). Wild: *Vicia* sp.

3. **Čavdar**, Sofia (Hopf 1973b). Early Neolithic (4800 bc). Scarce remains: einkorn wheat (rare); emmer wheat (rare); naked six-rowed barley (few); pea (prevailing); lentil (rare). Wild: *Pyrus/Malus*; *Quercus* acorns.

4. **Ovčarovo**, north-east Bulgaria (Janushevich 1978). Neolithic (Gumelnitsa culture, 3750 bc). Rich remains: einkorn wheat (prevailing); emmer wheat (frequent); free-threshing wheat (few); spelta wheat (rare); naked barley (frequent); bitter vetch (frequent); grass pea (rare).

5. **Nova Zagora**, Sliven (Hajnalová 1980). Bronze Age. Rich remains: einkorn wheat (frequent); emmer wheat (rare); naked six-rowed barley (frequent); pea (rare); lentil (rare); bitter vetch (prevailing). Wild: *Vicia* sp.; *Bromus* sp.; *Quercus* acorns.

Rumania

(General references: Wasylikowa *et al.* 1991; Cârciumaru 1991)

1. **Liubcova**, Caraş, Severin district (Cârciumaru 1991). Vinča culture B2. Rich remains: einkorn wheat (frequent); emmer wheat (frequent); free-threshing wheat (frequent); hulled and naked barley (frequent); lentil (prevailing).

2. **Cîrcea**, Dolj district (Cârciumaru 1991). Dudeşti culture (4500–4000 bc). Charred remains: free-threshing wheat (frequent); pea (a pure hoard of seeds). Impressions: einkorn wheat; barley.

3. **Cascioarele**, Calarasi district (Cârciumaru 1991; M. Hopf unpublished data). Gumelnitsa culture (*c.* 3500 bc). Rich remains: wheat; naked six-rowed barley (prevailing); hulled six-rowed barley (few); pea (few); a hoard of bitter vetch.

4. **Poduri**, Bacau district (Cârciumaru 1991) Pre-Cucuteni culture (1st half of 4th millennium bc). Hoards of charred grains of the following cereals, frequently contaminated by few grains of other cereals: einkorn wheat

(with some emmer wheat); emmer wheat (with some einkorn wheat); free-threshing wheat; hulled six-rowed barley; naked six-rowed barley.

5. **Sucidava – Celei**, Olt district (Wasylikowa *et al.* 1991). Transitional period between Eneolithic and Bronze Age (*c.* 2200 bc). Rich remains, unspecified quantities: einkorn wheat; emmer wheat; spelta wheat?; free-threshing wheat; hulled six-rowed barley; lentil; faba bean; bitter vetch; flax – both seeds and textiles. Wild: *Camelina sativa*; *Vitis vinifera*.

Moldavia and Ukraine

(General references: Janushevich 1984; Wasylikowa *et al.* 1991.)

1. **Sakharovka**, Moldavia (Janushevich 1984; and personal communication). Early Neolithic, Bug-Dniester culture (4700 bc). Rich remains: (both imprints and carbonized grain): emmer wheat (prevailing); spelta wheat (rare); naked barley (frequent); pea (rare), Wild: *Prunus insititia*; *P. spinosa*; *Cornus mas; Malus* sp.

2. **Other Early Neolithic settlements in Moldavia**: including Soroki and Ruptura (Janushevich 1975). Bug-Dniester culture (*c.* 4800 bc). Scarce remains (imprints): einkorn wheat (few); emmer wheat (prevailing); spelta wheat? (rare).

3. **Early Eneolithic settlements in Moldavia** (Janushevich 1975, 1976, 1978, 1984). Tripolye culture, early phase (3850–3650 bc). Scarce remains (imprints): The most common imprints are those of einkorn wheat, emmer wheat and hulled and naked barley. They are accompanied by a single find of free-threshing wheat (at Floreshti); questionable and rare appearance of spelta wheat; broomcorn millet (rare); pea (few); bitter vetch (few).

4. **Starye Kukoneshti** (Shcherbaki), northern Moldavia (Janushevich 1978, 1984). Eneolithic, Tripolye culture, middle phase. Rich remains (mainly carbonized): einkorn wheat (few); emmer wheat (prevailing); naked barley (few). No other plants mentioned.

5. **Several Eneolithic settlements in the Ukraine**, mainly in the area of Kiev (Janushevich 1978, table 1). Tripolye culture, middle phase. Scarce remains (imprints); einkorn wheat (few); emmer wheat and naked barley (prevailing); hulled barley (few).

Hungary

(General references: Tempír 1964; Hartyányi and Nováki 1975; Wasylikowa *et al.* 1991.)

1. **Pari-Altäcker dülö**, Tolna (Hartyányi and Nováki 1975). Danubian (Bandkeramik) culture. Rich remains: einkorn wheat (frequent); emmer

wheat (prevailing). Wild: *Bromus* sp.; *Polygonum convolvulus*; *Galium* sp.

2. **Dévaványa-Réhelyi gát**, Békés (Hartyányi and Nováki 1975). Middle Neolithic (4420 bc). Rich remains: einkorn wheat (few); emmer wheat (frequent); naked six-rowed barley (prevailing); lentil (rare). Wild: weeds.

3. **Szilhalom**, Berettyóújfalu, Haidu-Bihar (Hartyányi and Máthé 1979). Late Neolithic. Rich remains: einkorn wheat (frequent); emmer wheat (frequent); pea (prevailing). Wild: *Malus* sp.

4. **Tiszaalpár-Várdomb** (Hartyányi 1982). Middle Bronze Age (late Vatya culture). Rich remains: einkorn wheat (frequent); emmer wheat (frequent); free-threshing wheat (few); hulled six-rowed barley (frequent); naked six-rowed barley (few); pea (rare); lentil (frequent); grass pea (frequent); linseed (seed and a single capsule). Wild: *Bromus secalinus; B. arvensis; Agrostemma githago; Lithospermum officinale; Polygonum convolvulus; P. aviculare; Chenopodium album; Convolvulus arvensis*.

5. **Pákozd-Vár**, Fejér (Hartyányi and Nováki 1975). Middle Bronze Age (Vatya culture). Rich remains: einkorn wheat (frequent); emmer wheat (few); barley (rare); pea (prevailing); faba bean (rare); bitter vetch (rare). Wild: *Malus* sp.; *Bromus* sp.; *Agrostemma githago*.

6. **Pécs-Nagyárpád**, Baranya (Hartyányi and Nováki 1975). Middle Bronze Age (Vučedol-Zók culture). Rich remains: einkorn wheat (frequent); emmer wheat (frequent); free-threshing wheat (prevailing); hulled six-rowed barley (rare); pea (rare); lentil (rare). Wild: *Camelina sativa; Bromus secalinus* and sp.; *Lolium* sp.; *Agropyron repens; Polygonum convolvulus; Rumex crispus; R. acetosella; Chenopodium album; Ch. murale; Malva* sp.; *Convolvulus arvensis; Centaurea* sp.; *Agrimonia eupatoria; Verbascum* sp.

Austria

(General references: Werneck 1949, 1961; Küster 1991.)

1. **Eggendorf am Walde**, Lower Austria (Werneck 1949). Neolithic, Lengyel group. Rich remains: emmer wheat (prevailing); free-threshing wheat (frequent).

2. **Mondsee**, east of Salzburg, Upper Austria (Hofmann 1924). Late Neolithic, Rich remains: emmer wheat (prevailing); free-threshing wheat (frequent); hulled six-rowed barley (frequent); pea (rare). Wild: *Rosa canina; Cornus sanguinea; Malus sylvestris; Corylus avellana; Fagus sylvatica; Quercus* sp.; *Rhamnus frangula; Tilia grandifolia; Fraxinus excelsior; Taxus baccata*.

3. **Gusen/Berglitzl**, Langenstein, near Linz, Upper Austria (M. Hopf, unpublished data). (i) Early Bronze Age. Rich remains: einkorn wheat (prevailing; some samples pure); emmer wheat (frequent); spelta wheat

(frequent). (ii) Middle Bronze Age. Rich remains: einkorn and emmer wheat (few); spelta wheat (few); broomcorn millet (prevailing). Wild: *Rubus* sp.; *Sambucus* sp.; *Polygonum* sp.; Cruciferae.

4. **Burgschleinitz**, Vienna (Werneck 1949, 1961). Late Bronze Age. Rich remains: emmer wheat (few); free-threshing wheat (prevailing); six-rowed barley (frequent); lentil (few); faba bean (frequent); broomcorn millet (frequent). Wild: *Bromus secalinus*; *Chenopodium album*; *Polygonum* sp.; *Sinapis arvensis*; *Vicia angustifolia*; *Quercus* sp.

5. **Dürrnberg/Hallein**, Salzburg (Werneck 1949). Late Bronze Age – transition to Iron Age. Numerous remains: emmer wheat (few); six-rowed barley (few); broomcorn millet (frequent); Italian millet (frequent); fibres of flax and hemp.

Italy and Malta

(General reference: Hopf 1991a.)

1. **Grotta dell'Uzzo**, Sicily (Costantini 1989). (i) Mesolithic (9th–7th millennia bc). Scarce remains: wild: *Arbutus unedo*; legume seeds (probably *Lathyrus* or *Pisum*); *Quercus* acorns; two pips of *Vitis*. (ii) Early Neolithic (first half of 6th millennium bc). Scarce remains: (unspecified quantities): einkorn wheat; emmer wheat; lentil; barley (somewhat later). (iii) Middle Neolithic (4800–4500 bc). Few remains: einkorn wheat; emmer wheat (prevailing); free-threshing wheat; barley (relatively common); lentil; pea; bitter vetch; faba bean. Wild: *Amygdalus* sp.; *Olea europaea*; *Vitis vinifera*; *Ficus carica*.

2. **Rendina**, Foggia (Follieri 1977–82). (i) Early Neolithic Impressed Ware (Cardial) culture. Phase II (5160–4810 bc). Scarce remains: wheat and barley. (ii) Phase III (4490 bc). Numerous remains (unspecified quantities): einkorn wheat; emmer wheat; free-threshing wheat; spelta wheat?; hulled six-rowed barley; lentil?; faba bean type (rare).

3. **Passo di Corvo**, Foggia (Follieri 1973). Impressed Ware (Cardial) culture (4190 bc). Scarce remains: einkorn wheat (few); emmer wheat (few); free-threshing wheat (prevailing); hulled six-rowed barley (rare); faba bean type (rare). Wild: *Avena* sp. (rare); *Vicia* sp.; *Lathyrus* sp.?; *Quercus* sp.

4. **Torre Canne**, Puglia (Punzi 1968). Early Neolithic (Cardial) culture. Scarce remains: einkorn wheat (few); emmer wheat (few); barley (prevailing).

5. **Pienza**, Siena (Castelletti 1976). (i) Impressed Ware (Cardial) culture. Numerous remains: einkorn wheat (rare); emmer wheat (prevailing); free-threshing wheat (frequent); hulled barley (few); naked barley (few). Wild: *Avena* sp.; *Vitis sylvestris*; acorns. (ii) Middle Neolithic. Numerous remains: emmer wheat (rare); free-threshing wheat (prevailing); barley

(rare). Wild: *Vicia* sp.; *Sambucus* sp.; *Avena* sp.; acorns. (iii) Late Neolithic. Numerous remains: emmer wheat (rare); free-threshing wheat (prevailing). Wild: *Avena* sp.; acorns. (iv) Bronze Age. Numerous remains: einkorn wheat (few); emmer wheat (prevailing); free-threshing wheat (few); hulled barley (few); naked barley (few); broomcorn millet (few). Wild: *Avena* sp.; *Vitis sylvestris*; *Cornus mas; Sambucus* sp.; acorns.

6. **Monte Còvolo**, Villa Nuova sul Clisi, Brescia (Pals and Voorrips 1979). (i) Neolithic, Terramare culture. Scarce remains: einkorn wheat (rare); emmer wheat (prevailing); six-rowed barley (few). Wild: *Cornus mas*; *Malus sylvestris; Prunus avium; P. spinosa; Vitis sylvestris; Quercus* sp.; *Rosa* sp.; *Sambucus* sp.; *Rubus* sp.; *Physalis alkekengi*. (ii). White Ware Phase. Scarce remains: einkorn wheat (few); emmer wheat (prevailing); six-rowed barley (frequent); broomcorn millet (rare). Wild: *Prunus avium*; *Quercus* sp.; *Vitis sylvestris*; *Rosa* sp.; *Sambucus* sp.; *Rubus* sp.; *Physalis alkekengi*. (iii) Bell Beaker Phase. Numerous remains: einkorn wheat (few); emmer wheat (prevailing); six-rowed barley (frequent). Wild: *Cornus mas*; *Malus sylvestris; Prunus avium; P. spinosa; Quercus* sp.; *Vitis sylvestris*; *Sambucus* sp.; *Physalis alkekengi*. (iv) Early Bronze Age. Scarce remains: emmer wheat (prevailing); six-rowed barley (few). Wild: *Malus sylvestris*.

7. **Skorba**, Malta (Helbaek 1966c). Neolithic, Ghar Dalam Phase (3810–3225 bc). Rich remains: emmer wheat (prevailing); free-threshing wheat (rare); six-rowed barley (frequent); lentil (few).

Poland

(General references: Klichowska 1976; Wasylikowa *et al.* 1991.)

1. **Rzeszów** (Klichowska 1976; Willerding 1980, table 4). Danubian (Bandkeramik) culture. Numerous remains (grains and imprints, unspecified quantities): einkorn wheat; emmer wheat; free-threshing wheat.

2. **Nowa Huta-Mogiła**, Karków district (Głuża 1983). Neolithic (Lengyel culture, 3480 bc). Rich remains: einkorn wheat (frequent); emmer wheat (prevailing); naked six-rowed barley (frequent); broomcorn millet (few). Wild: several species of *Bromus* and *Chenopodium*; *Polygonum*; *Convolvulus*; *Solanum nigrum*; *Corylus avellana*; *Sambucus ebulus*.

3. **Zesławice**, Kraków district (Giżbert 1960). Neolithic, Radial Pottery culture (2nd half of the 3rd millennium bc). Rich remains: einkorn wheat (frequent); emmer wheat (frequent); rye (few); broomcorn millet (lumps of pure grains); poppy (few). Wild: *Bromus secalinus*; *Polygonum aviculare*; *P. lapathifolium*; *Galium aparine; Corylus avellana*.

4. **Ćmielów**, Tarnobrzeg district (Klichowska 1976). Funnel Beaker culture (2825–2665 bc). Rich remains: emmer wheat (prevailing); pea (few); linseed (frequent). Wild: *Setaria glauca; Berberis vulgaris; Taxus baccata*.

5. **Ksiacznice Wielkie**, Kielce district. (Giżbert 1966). Funnel Beaker culture (*c*. 2800 bc). Rich remains: einkorn wheat (frequent); emmer wheat (frequent); free-threshing wheat (frequent); spelta wheat (frequent). Wild: *Bromus* sp.; *Lolium* sp.; *Lithospermum officinale*; *Lithospermum arvense*.

6. **Szlachcin**, Poznań district (Klichowska 1966). Funnel Beaker culture. Scarce remains (mostly imprints): einkorn wheat; emmer wheat; free-threshing wheat; hulled six-rowed barley. Carbonized remains: pea (few); broomcorn millet (lumps of pure grains).

Czechoslovakia

(General references: Tempír 1966, 1979; Hajnalová 1989; Wasylikowa *et al.* 1991.)

1. **Třtice**, Rakovník, Bohemia (Tempír 1973, 1979). Danubian (Bandkeramik). Numerous remains: einkorn wheat (frequent); emmer wheat (frequent); pea (prevailing). Wild: numerous weeds.

2. **Bylany**, Kutná Hora, Bohemia (Tempír 1979). (i) Danubian (Bandkeramik). Numerous remains: einkorn wheat (few); emmer wheat (prevailing); broomcorn millet (few); pea (few); lentil (few). (ii) Funnel Beaker culture. Rich remains: einkorn wheat (prevailing); emmer wheat (few); free-threshing wheat (few); six-rowed hulled barley (frequent). Wild: several weeds.

3. **Mohelnice**, Litovel, Moravia (Kühn 1981 Opravil 1979, 1981). Danubian (Bandkeramik) culture (4300–4200 bc). Numerous remains: einkorn wheat (rare); emmer wheat (prevailing); free-threshing wheat (frequent); two-rowed barley (few); broomcorn millet (rare); pea or vetch (rare); flax cord/fibre (few). Wild: *Malus sylvestris*; *Corylus avellana*.

4. **Blatné** near Štrky, Bratislava district, Slovakia (Hajnalová 1989). (i) Danubian (Bandkeramik). Numerous remains: einkorn wheat (frequent); emmer wheat (prevailing); spelta wheat (few); pea (rare). (ii) Middle Neolithic, Zeliezovce group. Numerous remains: einkorn wheat (frequent); emmer wheat (prevailing); spelta wheat (few); hulled six-rowed barley (rare); pea (rare); lentil (few).

5. **Košice-Barca**, Slovakia (Tempír 1969). Early Bronze Age (Otomani culture). Rich remains: einkorn wheat? (rare); emmer wheat (prevailing); free-threshing wheat (few); six-rowed barley (frequent); pea (frequent); lentil (frequent), Wild: *Bromus secalinus; Agrostemma githago; Polygonum convolvulus*.

6. **Nitriansky Hrádok**, Nové Zámky, Slovakia (Tempír 1969; Kühn 1981). (i) Early Bronze Age. Rich remains: einkorn wheat (frequent); emmer wheat (prevailing); free-threshing wheat (frequent); six-rowed barley and two-rowed barley (frequent); oat (rare); rye (rare); pea

(few/pure); lentil (rare). Wild: *Avena fatua; Bromus secalinus; Lolium* sp.; *Vicia* sp.; *Agrostemma githago*; *Quercus robur*. (ii) Bronze Age. Numerous remains: einkorn wheat (rare); emmer wheat (rare); six-rowed barley (prevailing); pea?. Wild: *Bromus secalinus*; *B. arvensis; Lolium temulentum*; *Agrostemma githago*; various Cruciferae.

Switzerland

(General references: Heer 1865; Neuweiler 1905, 1935, 1946; Jacomet *et al*. 1991; Küster 1991.)

1. **Niederwil** (van Zeist and Casparie 1974). Neolithic lake shore settlement (3700–3625 bc). Rich waterlogged and carbonized remains: emmer wheat (few); free-threshing wheat (prevailing); naked six-rowed barley (frequent); broomcorn millet (few); linseed including capsules (frequent); poppy (frequent). Wild: *Malus sylvestris; Corylus avellana; Sambucus nigra; Rubus idaeus; R. fruticosa; Fragaria vesca; Quercus* sp.; *Fagus sylvatica; Prunus padus; P. spinosa; Cornus sanguinea; Viburnum opulus; V. lantana*.

2. **Seeberg**, Burgäschisee-Süd (Villaret-von Rochow 1967). Neolithic. Rich water logged and carbonized remains: einkorn wheat (rare); emmer wheat (rare); free-threshing wheat (prevailing); six-rowed hulled and naked barley (frequent); pea (rare); linseed including capsules (few); poppy (frequent). Wild: *Brassica campestris* (rich); *Quercus* sp.; *Malus sylvestris; Prunus spinosa; Corylus avellana*; *Cornus sanguinea; Sambucus ebulus; Rubus idaeus; R. fruticosa; Fragaria vesca; Rosa* sp.; *Solanum dulcamara; Pastinaca sativa; Physalis alkekengi; Polygonum aviculare; P. persicaria; Hordeum murinum*.

3. **Twann** Bielersee (Piening 1981). Late Neolithic (3300–3000 bc). (i) Cortaillod group. Rich waterlogged and carbonized remains (often pure): einkorn wheat (few); emmer wheat (few); free-threshing wheat (prevailing); hulled six-rowed barley (frequent); naked six-rowed barley (few); pea seeds and fragments of pods (frequent); linseed (few); poppy (few). Wild. *Malus* sp.; *Polygonum convolvulus; Chenopodium album; Lapsana communis; Rubus idaeus; R. fruticosus; Fragaria vesca; Sambucus* sp.; *Physalis alkekengi; Corylus avellana*. (ii) Horgen group. Rich remains: einkorn wheat (rare); emmer wheat (few); free-threshing wheat (frequent); hulled six-rowed barley (prevailing); linseed (few); poppy (few). Wild: *Polygonum convolvulus; Rubus idaeus; R. fruticosus; Fragaria vesca*.

4. **Zürich**, Lake shore settlements (Jacomet *et al*. 1989; 1991). Rich remains from several sites (both carbonized or excellently preserved waterlogged material) covering the various phases of the lake shore agricultural settlements: (i) Neolithic: Egolzwil, Cortaillod, Pfyn, Horgen, and Schnurkeramik cultures (4400–2500 bc): In most sites and beds (except

Pfyn culture) emmer wheat is the commonest cereal, followed by large quantities of naked and hulled barley and by free-threshing wheat. (The latter abound in Pfyn culture.) Einkorn wheat appears in smaller quantities yet also it is occasionally common. The other characteristic crop remains – from the Egolzwil culture (4400 bc) onward – are pea, flax (both seeds and capsules) and poppy. The finds produce a rich array of plants collected from the wild, particularly *Malus sylvestris*, *Corylus avellana*, *Prunus spinosa*, *Rubus idaeus*, *Rubus fruticosus*, *Fragaria vesca*, and remains of numerous weeds and local herbs. (ii) Bronze Age (1600–950 bc): remains come mainly from the site of Mozartstrasse; the finds produced the same spectrum of cultivated plants, with the addition of spelta wheat, broomcorn millet, foxtail millet, lentil and *Camelina*. From the Pfyn culture onward, the finds include also some remains of Mediterranean aromatic plants: celery *Apium graveolens*; dill *Anethum graveolens* and *Melissa officinalis*.

Germany

(General references: Knörzer 1991; Küster 1991.)

1. **Hienheim**, Kr. Kelheim (Bakels 1978). Danubian (Bandkeramik) culture (4300–3900 bc). Rich remains: einkorn wheat and emmer wheat (prevailing, apparently in equal proportions); pea (frequent); lentil (few); linseed (rare). Wild: *Corylus avellana;* numerous weeds.

2. **Heilbronn**, including Heilbronn-Böckingen, H.-Gross-Gartach, H. -Willsbach (Bertsch and Bertsch 1949; Mattes 1957) Danubian (Bandkeramik) culture. Rich remains: einkorn wheat (frequent); emmer wheat (frequent); hulled six-rowed barley (frequent); naked six-rowed barley (few); pea (few); lentil (frequent); linseed (rare).

3. **Aldenhovener Platte**: Langweiler, Lamersdorf, Bedburg-Garsdorf, Meckenheim and Rödingen (Knörzer 1973, 1974, 1979). Danubian (Bandkeramik) culture. (4400–4000 bc). Rich remains: emmer wheat and einkorn wheat (abundant); pea (few); lentil (rare); linseed (few; in one place very frequent). Wild: *Papaver setigerum* (few); *Bromus secalinus* (abundant); *Vicia* sp.; *Quercus* sp.; *Prunus insititia*; *P. spinosa*; *Malus* sp.; *Corylus avellana*; *Sambucus* sp.

4. **Dresden-Nickern** (Baumann and Schultze-Motel 1968). Danubian culture. Rich remains: emmer wheat (few); pea (a pure hoard).

5. **Dannau**, Oldenburg (Kroll 1981b). Middle Neolithic. Rich remains: einkorn wheat (rare); emmer wheat (frequent); naked six-rowed barley (prevailing). Wild: *Corylus avellana; Rubus idaeus; Sambucus nigra*.

6. **Eberdingen-Hochdorf**, Kr. Ludwigsburg (Küster 1983). Middle Neolithic. Schussenried group (3000 bc). Rich remains: einkorn wheat (almost as frequent as naked barley); emmer wheat (few); free-threshing wheat (few); naked six-rowed barley (prevailing); pea (few); linseed (few); poppy

(few). Wild: *Corylus avellana; Fragaria vesca; Malus* sp.; *Sambucus; Petroselinum* sp.

7. **Federseeried** (Blankenhorn and Hopf 1982). Late Neolithic, Aichbühl-Schussenried group. Rich remains: einkorn wheat (few); emmer wheat (frequent); free-threshing wheat (prevailing); spelta wheat (few); hulled six-rowed barley (few); naked six-rowed barley (rare); poppy (rare). Wild: *Corylus avellana; Malus* sp.; *Fragaria vesca; Rubus idaeus; R. fruticosus; Fagus sylvatica.*

Belgium

(General reference: Bakels 1991.)

1. **Aubechies**, and other contemporary sites (Bakels and Roussell 1985). Danubian (Bandkeramik). Numerous remains: emmer wheat (prevailing); einkorn wheat (frequent); flax (rare). Wild: numerous herbs.
2. **'La Bosse de la Tombe' à Givry** (Heim 1979). Roessen group (*c.* 3400 bc). Scarce remains: emmer wheat (few); free-threshing wheat (prevailing). Wild: *Polygonum hydropiper; Stellaria media; Corylus avellana* (frequent).

The Netherlands

(General references: van Zeist 1968–70; Bakels 1991.)

1. **Beek-Kerkeveld**, Suid-Limburg (Bakels 1979). Danubian culture. Rich remains: einkorn wheat (prevailing); emmer wheat (frequent); linseed (rare); poppy (rare). Wild: *Chenopodium album; Bromus secalinus; Polygonum convolvulus; Malus sylvestris; Prunus spinosa; Corylus avellana; Echinochloa crus-galli.*
2. **Geleen-Haesselderveld**, Suid-Limburg (Bakels 1979). Danubian culture. Numerous remains: einkorn wheat? (few); emmer wheat (few). Wild: *Chenopodium album; Bromus* sp.; *Polygonum convolvulus.*
3. **Aartswoud**, Noord-Holland (Pals 1984). Late Neolithic (*c.* 2000 bc). Rich remains: einkorn wheat (few); emmer wheat (frequent); free-threshing wheat (few); naked six-rowed barley (prevailing); linseed (few). Wild: *Malus* sp.; *Rubus fruticosus; Corylus avellana; Quercus* sp.
4. **Emmerhout 215**, Emmen (van Zeist 1968–70, p. 73). Bronze Age (1370 bc). Rich remains: einkorn wheat (numerous) naked wheat (prevailing). Wild: *Corylus avellana; Spergula arvensis; Polygonum convolvulus; P. lapathifolium.*

Denmark

(General reference: Jensen 1991.)

1. **Sarup**, S. W. Fyn (Jørgensen 1981). Funnel beaker culture (2630–2450

bc). Numerous remains: einkorn wheat (rare); emmer wheat (prevailing); free-threshing wheat (rare); naked six-rowed barley (few). Wild: *Malus sylvestris*; *Corylus avellana*; *Chenopodium album*.

2. **Bundsø**, Als (Jessen 1939). Funnel Beaker culture MN III. Numerous remains (charred and imprints): einkorn wheat (prevailing); emmer wheat (frequent); free-threshing wheat (few); naked six-rowed barley (frequent). Wild: *Malus sylvestris; Rubus idaeus; R. fruticosus; Corylus avellana*.

3. **Birknaes**, east Jutland (Helbaek 1952a; Jørgensen 1979). Late Neolithic (^{14}C = 1680–1570 bc; calendar years = 2070–1930 BC). Rich remains: einkorn wheat (frequent); emmer wheat (frequent); free-threshing wheat (frequent); spelta wheat (few); naked six-rowed barley (prevailing).

4. **Vadgaard**, north Jutland (Jørgensen 1979). Early Bronze Age II (^{14}C = 1250 bc; calendar years = 1550 BC). Scarce remains: spelta wheat (prevailing) (62 grains and 700 spikelets); hulled and naked six-rowed barley (few). Wild: *Bromus mollis; Polygonum aviculare; P. persicaria*.

5. **Nørre Sandegaard**, Bornholm (Helbaek 1952b). Bronze Age III. Rich remains: einkorn wheat (frequent); emmer wheat (prevailing); free-threshing wheat (rare); broomcorn millet (rare). Wild: *Malus sylvestris* (large quantities); *Corylus avellana*; *Crataegus* sp.; *Cenococcum geophilum*.

Sweden

(General references: Hjelmqvist 1979a; Jensen 1991.)

1. **Eker**, Hjulberga, Rosenland, Närke (Hjelmqvist 1979a). Early Neolithic. Scarce remains (carbonized): emmer wheat (rare); free-threshing wheat (few); naked six-rowed barley (prevailing); pea (rare). Wild: *Corylus avellana*; *Sinapis arvensis*. Imprints: 2 emmer wheat; 10 free-threshing wheat; 12 naked six rowed barley.

2. **Alvastra**, Östergötland (Hjelmqvist 1955). Middle Neolithic. (1500 bc). Numerous remains (carbonized): einkorn wheat (rare); hulled six-rowed barley (rare); naked six-rowed barley (prevailing). Wild: *Malus* sp. (fruits and pips); *Corylus avellana*. Imprints: 12 naked six-rowed barley; 2 hulled six-rowed barley.

Norway

(General reference: Jensen 1991.)

1. **Rugland**, Jaeren (Bakkevig 1982). *c*. 1500 bc. Few remains: charred six-rowed naked barley. The oldest cultivated material known from Norway.

Finland

(General reference: Jensen 1991.)

1. **Niuskula**, Turko, south-west Finland (Vuorela and Lempiäinen 1988). *c.* 1400 bc. Numerous remains: charred six-rowed naked barley. The oldest cultivated material known from Finland.

Great Britain

(General reference: Greig 1991.)

1. **Windmill Hill**, Avebury, Wilts and other Neolithic sites in south England (Helbaek 1952c). Neolithic (*c.* 3000 bc). Scarce remains (imprints): einkorn and emmer wheats (prevailing): hulled and naked barley (few); flax (rare). Wild: *Malus sylvestris* (few).

2. **Balbridie**, Kincardineshire, Grampian, Scotland (Boyd 1988). Neolithic (*c.* 2500 bc). Carbonized remains: emmer wheat, club wheat, naked six-row barley, flax; wild: *Malus sylvestris*, *Avena* sp.

3. **West Row Fen**, Suffolk (Martin and Murphy 1988). Early Bronze Age (*c.* 1400–1200 bc). Scarce carbonized remains: emmer wheat (prevailing); spelta wheat (few); hulled barley (few); flax: seeds and capsules (frequent). Wild: *Corylus avellana* and others.

France

(General references: Courtin and Erroux 1974; Erroux 1976; Marinval 1988; Hopf 1991a.)

1. **La Baume Fontbrégoua**, Salernes, Var (Courtin and Erroux 1974). (i) Mesolithic. Few remains: Wild: *Vicia* sp.; *Vitis sylvestris*. (ii) Impressed Ware (Cardial) culture (4700–4000 bc). Scarce remains: free-threshing wheat (prevailing). Wild: *Lathyrus* cf. *cicera*; *Vicia* sp.; *Quercus* sp. (iii) Middle Neolithic (3500–3000 bc). Rich remains: einkorn wheat (few); emmer wheat (prevailing); free-threshing wheat (frequent); six-rowed barley (few); naked six-rowed barley (few). Wild: *Vicia* sp.; *Lathyrus* cf. *cicera*.

2. **Châteauneuf-les-Martigues**, Font des Pigeons, Bouches-du-Rhône (Courtin *et al.* 1976). Impressed Ware (Cardial) culture (4120 and 4100 bc). Rich remains: free-threshing wheat and naked six-rowed barley in equal proportions. Wild: *Prunus cerasus*; *Pinus* sp.

3. **Kirschnaumen – Evendorff**, Moselle (Decker *et al.* 1977; Bakels

1984). Danubian. Scarce remains: emmer wheat (prevailing); pea (few); lentil (few).

4. **Baume de Gonvillars**, Gonvillars, Haute Saône (Courtin *et al.* 1976). Danubian (4300 bc). Rich remains: einkorn wheat (rare); emmer wheat (prevailing); free-threshing wheat (frequent); naked six-rowed barley (frequent).

5. **Menneville**, Aisne; and several other sites in the Aisne Valley (Bakels 1984). (i) Danubian (Bandkeramik). Few remains: emmer wheat (few); naked barley (more common). (ii) Middle Neolithic. Few remains: einkorn wheat (rare) emmer wheat (few); six-rowed barley (few); poppy (frequent).

6. **Grotte Antonnaire**, Montmaur en Diois, Drôme (Coquillat 1956; M. Hopf, unpublished data). Middle Neolithic. Numerous remains: einkorn wheat (rare); emmer wheat (few); free-threshing wheat (prevailing); six-rowed barley (few); naked six-rowed barley (frequent). Wild: *Avena* sp.; *Sambucus ebulus; Rubus idaeus.*

7. **Grotte G, Baudinard**, Var (Courtin and Erroux 1974). Chasséen. Numerous remains: emmer wheat (frequent); free-threshing wheat (frequent); six-rowed barley (prevailing); chickpea (rare); faba bean (rare). Wild: *Lathyrus cicera; Vicia* sp.; *Quercus* sp.

8. **Station III de Clairvaux**, Jura (Lundström-Baudais 1984). Late Neolithic lakeside settlement. Numerous mostly waterlogged remains of threshing and of wild plants: emmer wheat (frequent); free-threshing wheat (few); six-rowed barley (prevailing); flax: seed, capsules, and textile (few); pea (rare); poppy (few). Wild: numerous species including *Corylus avellana; Malus sylvestris; Prumus spinosa; Rubus idaeus; R. fruticosus; Fraqaria vesca.*

9. **Ouroux-Marnay**, Ouroux, Saône-et-Loire (Coquillat 1964; Bonnamour 1974; Hopf 1985). Late Bronze Age. Rich remains: einkorn wheat (few); emmer wheat (few); spelta wheat (few); free-threshing wheat (few); hulled and naked six-rowed barley (prevailing); broomcorn and foxtail millets (frequent); poppy (cultivated? few). Wild: *Avena* sp.; several weeds; *Vitis sylvestris; Quercus pedunculata; Crataegus monogyna; Viburnum lantana.*

Spain

(General references: Buxó 1985; Hopf 1991a.)

1. **Coveta de l'Or**, Beniarrés, Alicante (Hopf and Schubart 1965; Lopez 1980). Early Neolithic Impressed Ware (Cardial) culture (4670 and 4315 bc). Rich remains: einkorn wheat (few); emmer wheat (frequent); free-threshing wheat (prevailing); hulled six-rowed barley (few); naked six-rowed barley (frequent). Wild: *Quercus* sp.

2. **Cueva de los Murciélagos**, Córdoba (Hopf & Muñoz 1974). Early Neolithic (4300–4160 bc). Rich remains: emmer wheat (prevailing); free-threshing wheat (few); naked six-rowed barley (frequent).

3. **Cueva del Toro**, El Torcal, Antequera, Málaga (Hopf, unpublished data). Middle Neolithic. Scarce remains: emmer wheat (rare); free-threshing wheat (few); naked six-rowed barley (prevailing); lentil (few); faba bean (few). Wild: *Quercus* acorns; *Celtis australis.*

4. **Cueva de Nerja**, Málaga (Hopf and Pellicer Catalán 1970). Chalcolithic (3115 bc). Rich remains: emmer wheat and free-threshing wheat (frequent); naked six-rowed barley (prevailing). Wild: *Quercus* sp.; *Olea europaea.*

5. **El Argar**, Almeria, and several other sites of the El Argar culture (Stika 1988; Hopf 1991b). Rich remains: emmer wheat (few); free-threshing wheat (frequent); hulled and naked six-rowed barley (prevailing); pea (few); faba bean (few); flax – both seeds and fibres (frequent). Wild: *Olea europaea*; *Vitis vinifera*; *Ficus carica*; *Pistacia* sp.; *Quercus* acorns; *Celtis australis*; *Stipa tenacissima.*

Portugal

(General references: Pinto da Silva 1988; Hopf 1991a.)

1. **Pedra de Ouro**, Alenquer (Pinto da Silva 1988). Neolithic, Bell – Beaker culture. Few remains: free-threshing wheat; faba bean.

2. **Vila Nova de S. Pedro**, Cartaxo (do Paço 1954; Pinto da Silva 1988). (i) Chalcolithic (2400–2200 bc). Scarce remains: faba bean; *Quercus* acorns. (ii) Bronze Age (2000–1600 bc). Numerous remains: free-threshing wheat (prevailing); naked and hulled six-rowed barley (few); faba bean (frequent); pea (rare); linseed (few). Wild: *Vicia* sp.; *Prunus avium; Pinus pinea*; legumes.

3. **Zambujal**, Torres Vedras (Hopf 1981a). (i) Chalcolithic (2400–2200 bc). Scarce remains: einkorn/emmer wheat (rare); free-threshing wheat (few); hulled and naked six-rowed barley (few); faba bean (prevailing); olive (wild? (few)). Wild: *Pinus* sp.; *Quercus* sp. (ii) Bronze Age (2000–1600 bc). Numerous remains: einkorn/emmer wheat (rare); free-threshing wheat (frequent); hulled and naked six-rowed barley (frequent); faba bean (prevailing); flax (rare); olive (wild? frequent). Wild: *Pinus* sp.; *Quercus* sp. acorns and cork of *Quercus suber.*

4. **Castrum of Baiões**, Beira Alta, Viseu district (Pinto da Silva 1976). Atlantic Late Bronze Age. Numerous remains: free-threshing wheat (few); hulled six-rowed barley (frequent); naked six-rowed barley (rare); pea (frequent); faba bean (prevailing).

11
Conclusions

The main aim of this book is a review of the available information on the origin and spread of cultivated plants in West Asia and Europe. The evaluation has been based on two sources of evidence: first of all, information obtained by analysis of plant remains retrieved from archaeological excavations, where early archaeological contexts, namely Mesolithic, Neolithic, and Bronze Age cultures have been the main source, and second, evidence provided by living plants, particularly by the wild progenitors of the cultivated plants. The combined information from these two sources leads to the following conclusions.

Beginning of domestication

The first definite signs of plant cultivation in the Old World appear in a string of early Neolithic farming villages that developed in the Near East by 7600–7000 bc (Map 2). The initiation of food production in this 'nuclear area' is based on the domestication of a relatively small number (eight to nine species) of local grain plants. It is worth noting that the start of plant cultivation in the Near East came more or less together with the domestication of animals. Sheep and goats were brought under human control soon after the start of plant cultivation (Uerpmann 1989, and personal communication). They were followed by cattle and pig domestication.

The Neolithic Near East crop assemblage

The crops of early Neolithic agriculture in the Near East are fairly well recognized. The most numerous vegetable remains in early farming villages come from three cereals: emmer wheat (*Triticum turgidum* subsp. *dicoccum*), einkorn wheat (*T. monococcum*), and barley (*Hordeum vulgare*). Diagnostic morphological traits (non-brittle ears, broad kernels) traceable in the archaeological finds indicate that by 7000 bc these three annual grasses were intentionally sown and harvested in a string of Pre-Pottery Neolithic B sites in the Near East. Emmer wheat and barley seem to have been the more common crops. Einkorn wheat is somewhat less frequent.

Several grain legumes appear as constant companions of the cereals. The most frequent pulses in the early Neolithic Near East contexts are lentil (*Lens culinaris*) and pea (*Pisum sativum*). Two more local legume crops are bitter vetch (*Vicia ervilia*) and chickpea (*Cicer arietinum*). In contrast to the cereals, remains of pulses usually lack morphological features by which initial stages of domestication can be recognized. Clear indications of lentil cultivation appear at about 6800 bc; and of pea, chickpea, and bitter vetch at about 6000 bc. Probably all four legumes were taken into cultivation somewhat earlier, either together with wheats and barley or soon after the domestication of those cereals. The origin of a fifth important Old World pulse, namely the faba bean (*Vicia faba*), has not yet been satisfactorily clarified. But a discovery of a hoard of seed in a Pre-Pottery Neolithic B site in north Israel (p. 107), indicates that the faba bean may also have been a member of the early Neolithic Near East crop assemblage.

Finally, flax (*Linum usitatissimum*) belongs to the Near East group of founder crops. It is impossible to decide whether the material obtained from Early Neolithic layers represents flax collected from the wild or remains of cultivated forms. Yet like the legumes both direct and circumstantial evidence indicates that by 6000 bc flax was already cultivated in the Near East.

Wild progenitors

The wild ancestors of the majority of the food plants grown in west Asia, Europe, and the Nile Valley are already well identified. The distribution areas and the main ecological preferences of most of these wild progenitors are also well elucidated. Comparison of this evidence with the archaeological information reveals that with practically all early crops the first signs of cultivation appear in the same general areas where the wild ancestral stocks abound today.

The geographic distribution of the wild progenitors of the Neolithic grain crops is indeed significant. Apart from flax and barley, the wild ancestors of the founder crops have a rather limited distribution. Wild emmer wheat and wild chickpea are endemic to the Near East arc. Assuming that their distribution did not change drastically during the last 10 millennia, the domestication of these crops could only have taken place in this restricted area. Because cultivated emmer wheat appears to be the most important Neolithic crop all over south-west Asia, Europe, and Egypt, the confinement of its wild progenitor to the Near East 'arc' delimits not only the place of origin of this cultivated cereal. It also marks the rather restricted geographic area where Old World Neolithic agriculture could have originated. Wild forms of einkorn wheat, lentil, pea, and bitter vetch have a somewhat wider distribution, but all, including barley, are centred in the Near East; that is, the region in which the earliest farming villages have been discovered.

The spread of the Near East crops

A most remarkable feature of Near East Neolithic agriculture is the rapid expansion it underwent soon after its establishment in the nuclear area. The quality and quantity of the archaeobotanical evidence varies considerably from region to region. Comprehensive information is available from most parts of Europe, but there is much sparser and frequently incomplete documentation from Caucasia, central Asia, and the Indian subcontinent. In Africa, critical data on plant remains are available only from Egypt. In spite of such uneven documentation, the following main features of the diffusion seem apparent.

The spread of agriculture from its Near East nuclear area to Europe and central Asia involves the Neolithic crop assemblage. Map 20 summarizes the information on the six most important Near East crops: emmer wheat (including its free-threshing derivatives), einkorn wheat, barley, lentil, pea, and flax. From the data assembled in this map and in Chapter 10 it is evident that the crops domesticated in the Near East nuclear area were also the initiators of food production in Europe, central Asia, the Indus basin, and the Nile Valley. The earliest farming cultures all over these vast regions always contain wheat and barley, and one, two, or more other Near East founder crops are frequently present in the sites as well.

The spread of the Near East crop assemblage both westwards (to Europe) and eastwards (to central Asia and to the Indian subcontinent) was a quick process (see Map 20). At the beginning of the 6th millennium bc, agriculture had already appeared in Greece. By the end of the 6th millennium bc, these crops were grown in Starčevo in the Danubian basin, in Merimde in the Nile Valley, in Chokh in the Caspian Sea belt, in Djeitun in Turkmenia, and in Mehrgarh in Pakistan. Less than 800 years later the Danubian (Bandkeramik) culture was already firmly established in loess soil regions all over central Europe, extending west into north France. At more or less the same time, and perhaps somewhat earlier, farming villages of the Impressed Ware (Cardial) culture appeared on the shores of the Mediterranean Sea and reached south Spain.

Substantial information on the age and spread of early farming cultures is available from Europe, where radiocarbon dating of sites representing the beginning of farming already permits the reconstruction of the diffusion. The evidence from central Asia and the Indian subcontinent is much more fragmentary. Yet the finds retrieved from Mehrgarh (p. 211), Chokh (p. 210) or Djeitun (p. 211) demonstrate that the diffusion of the Near East crops towards Transcaucasia, central Asia and the Indian subcontinent happened relatively early. All over those vast areas the start of food production depended on the same Near Eastern crops.

Information on archaeological sites which appear on Map 20.

1: Ali Kosh (Helbaek 1969). **2:** Jarmo (Helbaek 1959b). **3:** Tell es-Sawwan (Helbaek 1964b). **4:** Tell Bouqras (van Zeist and Waterbolk-Rooijen 1985). **5:** Yarym Tepe (Bakhteyev and Yanushevich 1980). **6:** Çayönü (van Zeist 1972). **7:** Can Hasan (French *et al.* 1972). **8:** Çatal Hüyük (Helbaek 1964a). **9:** Erbaba (van Zeist and Buitenhuis 1983). **10:** Hacilar (Helbaek 1970). **11:** Andreas Kastros (van Zeist 1981). **12:** Khirokitia (Waines and Stanley Price 1975–77). **13:** Tell Abu Hureyra (Hillman 1975). **14:** Ras Shamra (van Zeist and Bakker-Heeres 1986a). **15:** Ramad, and **16:** Tell Aswad (van Zeist and Bakker-Heeres 1985). **17:** Jericho (Hopf 1983). **18:** Ain Ghazal (Rollefson and Simmons 1985). **19:** Beidha (Helbaek 1966a). **20:** Merimde (Stemler 1980; M. Hopf, unpublished). **21:** Fayum (Caton-Thompson and Gardner 1934; Stemler 1980). **22:** Arukhlo (Lisitsina 1984; Janushevich 1984). **23:** Chokh (Lisitsina 1984; Schultze Motel 1989). **24:** Djeitun (Hillman and Charles, in press). **25:** Altyn Tepe (Janushevich 1984). **26:** Tepe Yahya (Costantini and Costantini-Biasini 1985). **27:** Mehrgarh (Jarrige and Meadow 1980; Costantini 1992). **28:** Harappa and several other Harappan sites (Vishnu-Mittre and Savithri 1982; Kajale 1991). **29:** Knossos (Renfrew 1979). **30:** Franchthi cave (Hansen 1992). **31:** Sesklo (Kroll 1981a). **32:** Ghediki (Renfrew 1966). **33:** Argissa (Hopf 1962; Kroll 1981a). **34:** Nea Nikomedeia (van Zeist and Bottema 1971). **35:** Anza (Renfrew 1976). **36:** Obre (Renfrew 1979). **37:** Gomolava (van Zeist 1975). **38:** Starčevo (Renfrew 1979). **39:** Čavdar (Hopf 1973b). **40:** Kazanluk (Renfrew 1979). **41:** Karanovo. and **42:** Azmaška (Hopf 1973b). **43:** Ovčarovo (Janushevich 1978). **44:** Liubcova, **45:** Cîrcea, and **46:** Bălăneasa (Cârciumaru 1991). **47:** Soroki, Novye Ruseschty, and other settlements in Moldavia (Janushevich 1975). **48:** Sakharovka (Janushevich 1984). **49:** Starye Kukoneshti (Janushevich 1978, 1984). **50:** Eneolithic settlements in the Ukraine (Janushevich 1978). **51:** Pari-Altäcker, **52:** Zánka, and **53:** Dévaványa (Hartyányi and Nováki 1975). **54:** Eggendorf (Werneck 1949). **55:** Mondsee (Hofmann 1924). **56:** Grotta dell 'Uzzo (Costantini 1989). **57:** Passo di Corvo and Rendina (Follieri 1973, 1977–82). **58:** Torre Canne (Punzi 1968). **59:** Pienza (Castelletti 1976). **60:** Monte Còvolo (Pals and Voorrips 1979). **61:** Skorba, Malta (Helbaek 1966c). **62:** Rzeszów, and **63:** Ojców/Kraków (Klichowska 1976). **64:** Nowa Huta (Gluza 1983). **65:** Strzelce, and **66:** Pietrowice (Klichowska 1976). **67:** Danubian settlements in Košice and Prešov areas (Willerding 1980). **68:** Mohelnice (Kühn 1981). **69:** Bylany, and **70:** Trtice (Tempír 1979). **71:** Niederwil (van Zeist and Casparie 1974). **72:** Zürich (Jacomet *et al.* 1989, 1991). **73:** Seeberg (Villaret-von Rochow 1967). **74:** Heilbronn, including Böckingen and Gross-Gartach (Bertsch and Bertsch 1949). **75:** Hienheim (Bakels 1973). **76:** Eisenberg (Rothmaler and Natho 1957). **77:** Dresden (Baumann and Schultze-Motel 1968). **78:** Helmstedt and Eitzum (Hopf 1963). **79:** Göttingen including: Gieboldehausen, Euzenberg, and Rosdorf (Willerding 1980). **80:** Aldenhoven, including Langweiler, Lamersdorf, Bedburg-Garsdorf, Meckenheim and Rödingen (Knörzer 1979, 1991). **81:** Sarup (Jørgensen 1981). **82:** Eker (Hjelmqvist 1979a). **83:** Alvastra (Hjelmqvist 1955). **84:** Rugland (Bakkevig 1982). **85:** Niuskula (Vuorela and Lempiäinen 1988). **86:** Sittard, Beek-Kerkeveld, and Gelee-Haesselderveld (Bakels 1979). **87:** La Bosse de la Tombe à Givry (Heim 1979). **88:** Windmill Hill (Helbaek 1952c). **89:** Kirschnaumen-Evendorff (Decker *et al.* 1977; Bakels 1984). **90:** Baume de Gonvillars (Pétrequin 1974). **91:** Fontbrégoua (Courtin and Erroux 1974). **92:** Chateauneuf-les-Martigues (Courtin *et al.* 1976). **93:** Grotte d'Aigle (Erroux 1979). **94:** Menneville (Bakels 1984). **95:** Perte du Cros, Saillac (Hopf 1967). **96:** Coveta de l'Or (Hopf and Schubart 1965; Lopez 1980). **97:** Cueva de los Murciélagos, Córdoba (Hopf and Muñoz 1974). **98:** El Argar and Almizarque (Hopf 1991b). **99:** Zambujal (Hopf 1981a). **100:** Vila Nova de S. Pedro (do Paço 1954).

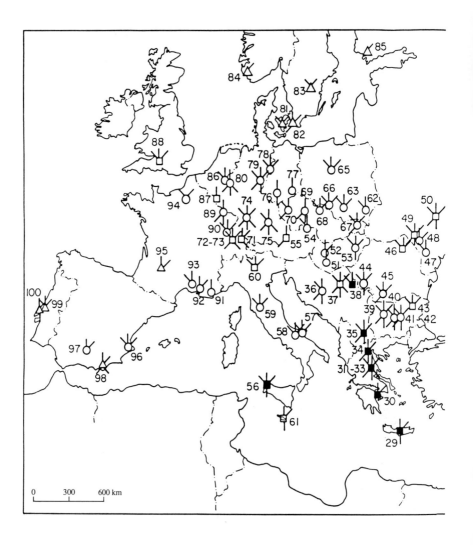

Map 20. The spread of the Near East Neolithic crop assemblage in Europe, west Asia, and north Africa. For details on the numbered sites, see p.231.

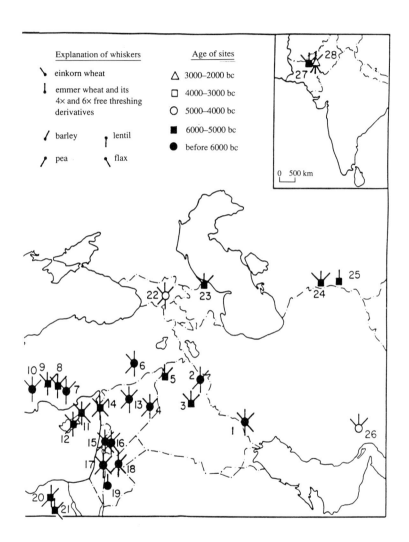

Availability of the archaeological evidence

Any attempt to reconstruct the beginning and diffusion of agriculture in the Old World has to take into account the unevenness of the archaeological record. As stressed on p.vi, plant remains retrieved from Europe, south west Asia, and the Nile Valley provide us with a reasonable overview of the beginning of agriculture in these parts of the world, but the archaeological evidence available from east Asia and the Indian subcontinent is much less complete, and it is only just emerging from Africa south of the Sahara. Consequently while the start of food production in the Near East is relatively amply documented, the founder crops are adequately identified, and the expansion to Europe and west Asia convincingly elucidated, there are discouragingly few solid facts on the beginning of plant cultivation in east Asia and in Africa. The time and place of origin of cultivated rice, common millet, sorghum, Old World cottons, and numerous other African and south-and east-Asiatic crops is only partly understood, because although most of their wild progenitors are already identified, they are unfortunately spread over wide areas. The evidence from the living plants therefore provides only a rough delimitation of where these plants could have been brought into cultivation. Until satisfactory archaeological and botanical evidence becomes available, the knowledge of the origin of south Asian, east Asian, and African staples will necessarily remain inadequate.

At present, then, our picture of crop plant evolution in Eurasia and Africa is obviously skewed. While there is relatively sound information on the developments in the classical Old World, we are still largely ignorant of the early events south and east of this area and we also know very little about the early interactions between west Asia and the major agricultural provinces in east Asia and in Africa south of the Sahara.

Early domestication outside the 'nuclear area'

Signs of additional domesticants start to appear soon after the introduction of Near East agriculture to Europe, central Asia, and the Nile Valley. The addition of some of these crops obviously took place outside the Near East but within the already established cultivation of the Near East crop assemblage. The poppy, *Papaver somniferum*, provides a well-documented example for such domestication. Both the area of distribution of the wild poppy and the archaeological finds (p. 130) indicate that *P. somniferum* was brought into cultivation in west Europe. It was locally added to the Near East grain-crop assemblage after its establishment in west Europe. Chufa, *Cyperus esculentus*, provides a similar example of an early local addition, this time in the Nile Valley (p. 186). Its dry tubers were found in large quantities in Egypt from pre-dynastic times onwards. The early appearance of the common millet, *Panicum miliaceum* in the Caspian basin

(p. 80), might indicate another local addition, but since the archaeological evidence from central and east Asia is still inadequate, it is impossible to decide whether *P. miliaceum* was locally added to the expanding Near East agriculture after it reached central Asia, or whether this cereal represents an east Asiatic domestication independent of the Near East diffusion.

Beginning and spread of horticulture

Olive, grape vine, fig, and date palm seem to have been the first principal fruit crops domesticated in the Old World. Definite signs of olive and date palm cultivation appear in Chalcolithic Palestine about 3700–3500 bc. Indications of date palm domestication are also available from contemporary lower Mesopotamia. We still do not know how extensive the Chalcolithic horticulture was. Except for Palestine, the archaeobotanical information on 4th millennium bc sites in other parts of the Levant is still grossly insufficient. The picture changes drastically in the Early Bronze Age (first half of the 3rd millennium bc). From this time on olives, grapes, and figs emerge as important additions to grain agriculture first in the Levant and soon later in Greece. They are subsequently planted all over the Mediterranean basin. The extensive cultivation of olives and grapes in the Bronze Age is reflected by the appearance of numerous presses and remains of storage facilities for olive oil and for wine. At the same time, dates were cultivated on the southern fringes and the warm river basins of the Near East arc, and during the New Kingdom they abound in the Nile Valley.

Apple, pear, plum, and cherry seem to have been added to the Old World horticulture much later, as definite signs of their cultivation appear only in the 1st millennium bc. Their culture is almost entirely based on grafting, so they could be domesticated and extensively cultivated only after the introduction of this sophisticated method of vegetative propagation.

Remains of fruit trees rarely show diagnostic anatomical traits by which one can distinguish between fruits collected from the wild or material harvested from cultivated orchards. To a large extent recognizing the domestication in fruit crops is based on circumstantial evidence such as the presence of fruit remains in areas in which the wild forms do not occur or on the quantitative analysis of artefacts associated with preparation of fruit products (e.g. oil, wine). This means that in fruit crops it is difficult to determine the initial stages of domestication, or in other words, it might well be that olive, grape, fig, or date cultivation did not start in the Chalcolithic (4th millennium bc) but had already begun in the late Neolithic (5th millennium bc).

In spite of these uncertainties, the following statements can be made: (a) The earliest definite signs of fruit tree cultivation appear in the Near East. (b) Horticulture developed only after the firm establishment of grain agriculture. (c) Like grain crops, several local wild fruits were taken into cultivation about the same time. (d) Domestication of fruit crops relied heavily on the invention of vegetative propagation. (e) Planting of perennial fruit-trees is a long-term investment, promoting a fully settled, sedentary way of life. (f) Soon after its successful establishment, horticulture spread from its original 'core area' into new territories in the Mediterranean basin and south-west Asia. (g) After the introduction of grafting (p. 135), the domestication of a whole group of 'second wave' fruit crops became possible.

The available archaeobotanical evidence on the beginning of fruit crop domestication is also supported by information on the wild relatives. Wild olive, grape vine, fig, and date are widely distributed over the Mediterranean and south-west Asia. They have a wide geographic distribution, so they do not provide critical values for a precise delimitation of the place of origin of these respective crops, but it is reassuring to know that forms from which the cultivated clones could have been derived thrive in wild niches in the east Mediterranean basin, and so the evidence from the living plants complements the archaeological finds. Most probably olive, grape vine, date, fig, as well as pomegranate and almond, were first brought into cultivation in the same *general area* which several millennia earlier saw the successful establishment of grain agriculture in the Old World. Chalcolithic and Bronze Age cultures in the east Mediterranean basin are characterized not only by the use of copper and bronze; 4th-millennium bc human societies in that region had also mastered horticulture.

Vegetables

This is the least-known group among the cultivated food plants of the Old World. Vegetable material consists almost entirely of perishable soft tissues, and such elements stand a meagre chance of charring and thus of surviving in archaeological contexts (p. 181). Consequently, only scanty remains of vegetables have been detected in archaeological excavations. The only exception is Egypt. In this especially arid country vegetables placed in pyramids and graves commonly survived by desiccation, and show that garlic, leek, onion, lettuce, melon, watermelon, and chufa were cultivated in the Nile Valley in the 2nd and the 1st millennia bc. As amply described by Keimer (1924, 1984), vegetable gardens constituted an important element of food production in Egyptian dynastic times.

Outside Egypt there are almost no early archaeobotanical finds of vegetable crops. However, early literary sources show that by the start of the 2nd millennium bc vegetable gardens flourished not only in the Nile Valley but also in Mesopotamia. Furthermore, in both areas the crops grown were more or less the same. The only major exception was chufa, which was restricted, almost entirely, to Egypt.

To sum up, the available evidence make it clear that by Bronze Age vegetable crops were part of food production both in Lower Mesopotamia and in Egypt. Very likely this geographic pattern is not accidental. In both regions we are faced with dense human settlement of very arid environments. In both survival depends on utilization of limited areas of irrigated or flooded land – bordered by large barren deserts. The latter have very little to offer in terms of a supplementary resource of green wild plants. Such shortage invites compensation. The early development of vegetable gardens might have been caused by such needs.

Weeds and crops

Several Old World grain plants, oil producers and vegetables seem to be 'secondary crops', that is, they first evolved as weeds and only later were picked up as crops (p.10). Oat, *Avena sativa*, and false flax, *Camelina sativa*, are well documented examples of this mode of evolution under domestication. It is also very likely that radish, lettuce, beet, leek, and several other vegetables entered cultivation through the same 'back door'. The incorporation of secondary crops into Old World food production seems to have happened rather late, since definite signs of their cultivation appear in Europe and west Asia only in the 2nd and 1st millennia bc.

Latecomers from other regions

A major wave of new crops arrived to the Near East in the 1st millennium bc, bringing into this region both Indian and African elements.

Warm weather crops make up the largest group among these immigrants, and include sorghum, sesame, rice, and Old World cotton. These crops started in the Mediterranean countries the tradition of summer crops, which from classical times, became an integral element of food production in these territories. Archaeobotanical evidence indicates that rice, sorghum, sesame, and cotton were already part of Indian agriculture in the 2nd millennium bc. All appear in Harappan sites. However, while rice, *Oryza sativa*, and sesame, *Sesamum indicum*, are south- and east-Asiatic elements, sorghum is not. Very likely the latter crop was taken into cultivation in east Africa. It was first introduced to the Indian subcontinent and only from there it was eventually transported to the Near East.

A second group of latecomers from the east comprises several fruit trees, prominent among which are apricot and peach. Both are cool climate elements that appeared in the Near East and the Mediterranean basin only at the end of the 1st millennium bc. Another, even earlier, immigrant is the citron, which was the first representative of south-east Asiatic citrus fruits to arrive in the Mediterranean basin.

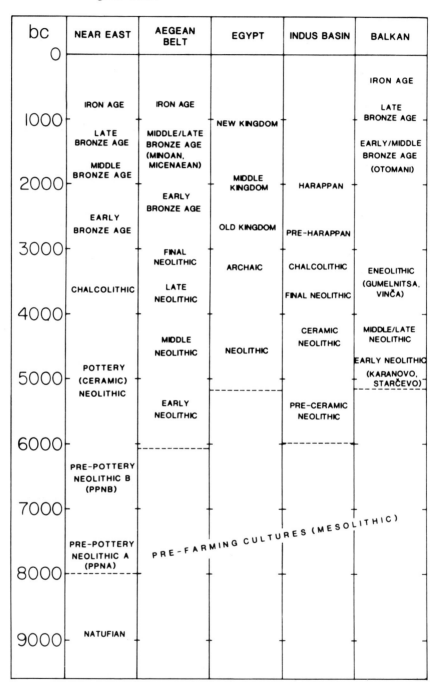

Fig. 45. Chronological chart for the main geographical regions.

WEST MEDITER- RANEAN	CENTRAL EUROPE	UKRAINE	ALPINE BELT	SCANDINAVIA	bc
					0
				IRON AGE	
	IRON AGE	IRON AGE			
					1000
				BRONZE AGE	
EARLY/MIDDLE BRONZE AGE (EL ARGAR)	BRONZE AGE	BRONZE AGE	BRONZE AGE		
				LATE NEOLITHIC	2000
CHALCOLITHIC			LATE NEOLITHIC	EARLY NEOLITHIC (FUNNEL BEAKER)	
	MIDDLE/LATE NEOLITHIC (RÖSSEN, LENGYEL, FUNNEL BEAKER)				3000
MIDDLE/LATE NEOLITHIC (CHASSÉEN)		ENEOLITHIC (TRIPOLYE)	EARLY NEOLITHIC LAKE SHORE SETTLEMENTS		
	EARLY NEOLITHIC (DANUBIAN = BANDKERAMIK)	EARLY NEOLITHIC			4000
EARLY NEOLITHIC (CARDIAL)					
					5000
PRE-FARMING CULTURES (MESOLITHIC)					
					6000
					7000
					8000
					9000

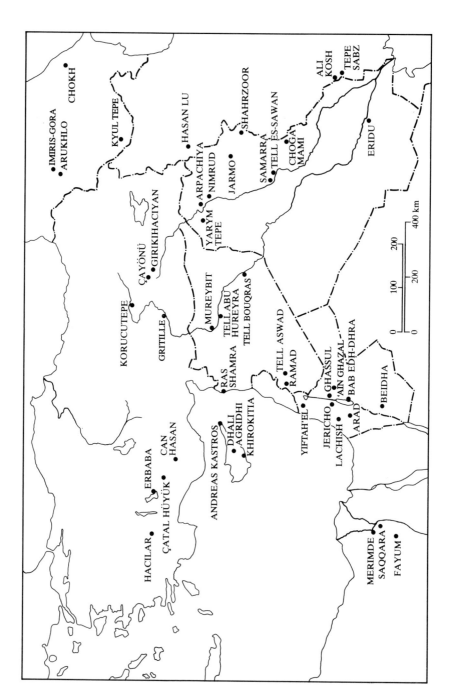

Map 21. The near East showing the locations of the main archaeological sites mentioned in this book.

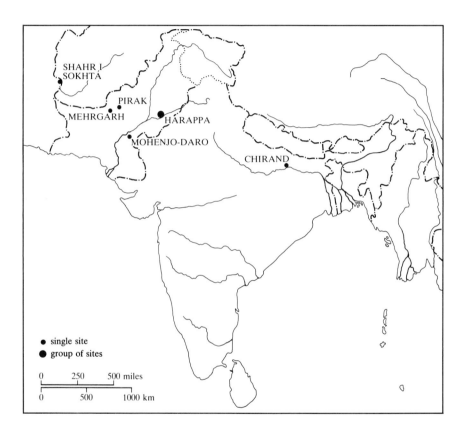

Map 22. The Indian subcontinent showing the locations of the main archaeological sites mentioned in this book.

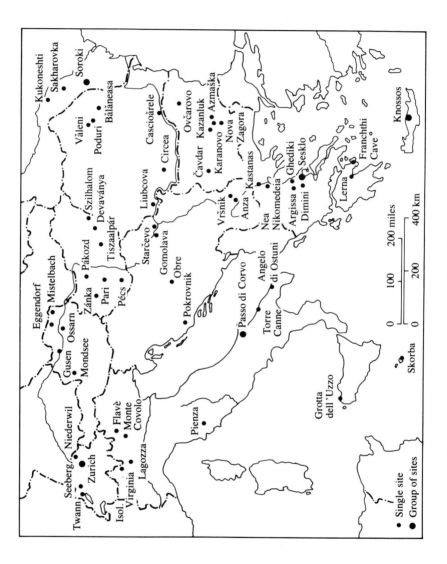

Map 23. South and south-east Europe showing the locations of the main archaeological sites mentioned in this book.

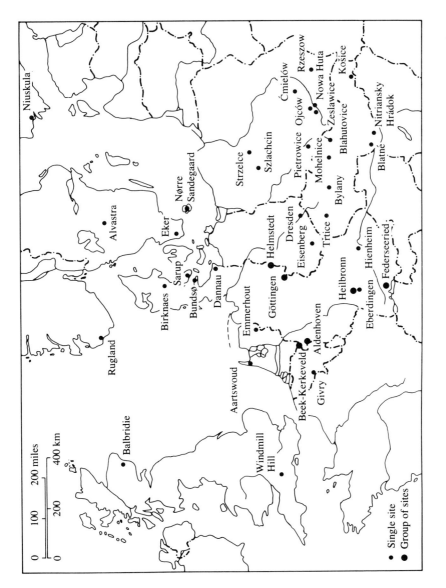

Map 24. Central and north Europe showing the locations of the main archaeological sites mentioned in this book.

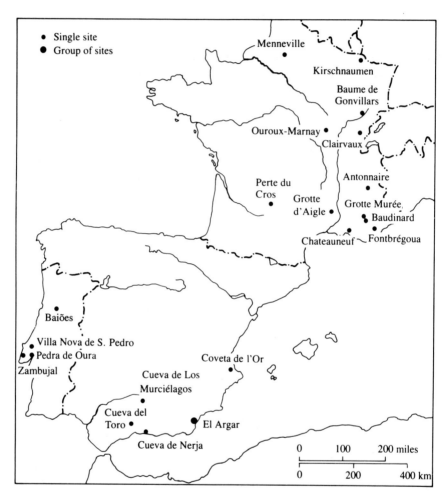

Map 25. France, Spain, and Portugal showing the locations of the main archaeological sites mentioned in this book.

References

Aaronsohn, A. (1910). *Agricultural and botanical exploration in Palestine*. 66 pp. Bureau Plant Industry USA, Bull. No. 180.

Allachin, F. R. (1982). Antecedents of the Indus civilization. *Proc. Br. Acad. Lond.* **66**, 135–60.

An, Z. (1989). Prehistoric agriculture in China. In: *Foraging and farming: the evolution of plant exploitation*. (eds. D.R. Harris and G.C. Hillman) pp. 643–9. Unwin & Hyman, London. New Jersey.

Baas, J. (1979). Kultur- und Nutzpflanzen aus einer römerzeitlichen Grube in Butzbach und ihr Zusammenhang mit Pflanzenfunden aus anderen römischen Fundstätten. *Saalburg-Jahrbuch*, Vol. 36, pp. 45–82. Walter de Gruyter, Berlin/New York.

Badler, V. R., McGovern, P.E., and Michel, R.H. (1990). Drink and be merry! Infrared spectroscopy and ancient Near Eastern wine. *Museum Applied Science Center for Archaeology (MASCA) Research Papers* Vol. 7. pp. 25–36. The University Museum of Archaeology and Anthropology, University of Pennsylvania, Philadelphia.

Bailey, C. H. and Hough, L. F. (1975). Apricots. In: *Advances in fruit breeding* (eds. J. Janick and J. N. Moore), pp. 367–83. Purdue University Press, West Lafayette, Indiana.

Bakels, C. C. (1978). Four Linearbandkeramik settlements and their environment. A paleoecological study of Sittard, Stein, Elsloo, and Hienheim. *Analecta Praehist. Leidensia* **11**, 1–248.

Bakels, C. C. (1979). Linearbandkeramische Früchte und Samen aus den Niederlanden. *Archaeo-Physika* **8**, 1–10.

Bakels, C. C. (1982). Der Mohn, die Linearbandkeramik, und das westliche Mittelmeergebiet. *Archäologisches Korrespondenzblatt* **12**, 11–13.

Bakels, C. C. (1984). Carbonized seeds from Northern France. *Analecta Praehist. Leidensia* **17**, 1–27.

Bakels, C. C. (1991). Western continental Europe. In: *Progress in Old World palaeoethnobotany* (eds. W. van Zeist, K. Wasylikowa, and K-E. Behre), pp. 279–98. Balkema, Rotterdam.

Bakels, C. C. and Rousselle, R. (1985). Restes botaniques et agriculture du Néolithique ancien en Belgique et aux Pays-Bas. *Helinum* **25**, 37–57.

Bakhteyev, F. Kh. and Yanushevich, Z. V. (1980). Discoveries of cultivated plants in the early farming settlements of Yarym-Tepe I and Yarym-Tepe II in northern Iraq. *J. Archaeol. Sci.* **7**, 167–78.

Bakkevig, S. (1982). Økologi og økonomi for deler av Sør-Jaeren i sen-neolitikum. Del 2. Makrofossil analyse. Saltvannsflotasjon av materiale fra Rugland på Jaeren. *AmS-Skrifter* **9**, 33–40.

Bar-Adon, P. (1980). *The cave of the treasure. The finds from the caves in Nahal Mishmar*. 243 pp. Israel Exploration Society, Jerusalem.

Baruch, U. (1990). Palynological evidence of human impact on the vegetation as recorded in Late Holocene lake sediments in Israel. In: *Man's role in the shaping of the eastern Mediterranean landscape* (eds. S. Bottema, G. Entjes-Nieborg, and W. van Zeist), pp. 283–93. Balkema, Rotterdam.

Bar-Yosef, O. and Alon, D. (1988). Nahal Hemar cave: the excavations. '*Atiqot* (Dept. Antiquities & Museums, Jerusalem) **38**, 1–30.

Baumann, W. and Schultze-Motel, J. (1968). Neolithische Kulturpflanzenreste aus Sachsen. *Arbeits- u. Forschungsberichte der sächsischen Bodendenkmalpflege* **18**, 9–28.

Bedigian, D. (1985). Is še-giš-ì sesame or flax? *Bulletin on Sumerian agriculture* (eds. J. N. Postgate and M. A. Powell) Vol. 2, pp. 159–78. Cambridge University Press.

Bedigian, D. and Harlan, J. R. (1986). Evidence for cultivation of sesame in the ancient world. *Econ. Bot.* **40**, 137–54.

Ben-Tor, A. (1975). The first season of excavation at Tell-Yarmuth. *Qedem* **1**, 54–87.

Ben Ze'ev, N. and Zohary, D. (1973). Species relationships in the genus *Pisum* L. *Israel J. Bot.* **22**, 73–91.

Beridze, R.K. and Kvatchadze, M.V. (1981). Origin and evolution of cultivated plums of Georgia, *Kulturpflanze* **29**, 147–50.

Bertsch, K. and Bertsch, F. (1949). *Geschichte unserer Kulturpflanzen* (2nd edn), 275 pp. Stuttgart.

Biagi, P. and Nisbet, R. (1987). Ursprung der Landwirtschaft in Norditalien. *Z. Archäol.* **21**, 11–24.

Blankenhorn, B. and Hopf, M. (1982). Pflanzenreste aus spätneolithischen Moorsiedlungen des Federseerieds. *Jahrb. Röm.-German. Zentralmus. Mainz* **29**, 74–99.

Boardman, J. (1976). The olive in the Mediterranean: its culture and use. *Phil. Trans. R. Soc. Lond., Biol. Sci.* **275**, 187–96.

Bond, D. A., Lawes, D. A., Hawting, G. C., Saxana, M. C., and Stephans, J. H. (1985). Faba bean. In: *Grain legume crops* (eds. R.J. Summerfield and E. H. Roberts), pp. 199–265. Collins, London.

Bonnamour, L. (1974). Trouvailles de la fin de l'âge du Bronze dans la Saône, sur le site d'Ouroux-Marnay (Saône-et-Loire). *Bull. Soc. Préhist. Franç.* **71**, 185–91.

Bor, N.L. (1968). *Triticum*. In: *Flora of Iraq* Vol. 9 (eds. C. C. Townsend, E. Guest, and A. Al-Rawi), pp. 194–208. Ministry of Agriculture, Baghdad, Iraq.

Bothmer, R. von, Jacobsen, N., Jørgensen, R. B., and Linde-Laursen, I. (1991). *An ecogeographical study of the genus Hordeum.* Systematic and Ecogeographic Studies on Crop Genepools 7. International Board for Plant Genetic Resources, Rome. 127 pp.

Bottema, S. (1980). On the history of the walnut in south eastern Europe. *Acta Bot. Neerl.* **29**, 343–9.

Bottema, S. and Woldring, H. (1984). Late Quaternary vegetation and climate of southwestern Turkey. Part II. *Palaeohistoria* **26**, 343–9.

Bowman, S. (1990). *Radiocarbon dating.* British Museum Publications, London. 64 pp.

Boyd, W.E. (1988). Cereals in Scottish antiquity. *Circaea* **5**, 101–10.

Braidwood, R. J. (1960). The agricultural revolution. *Sci. Am.* **203**, 131–48.

Braun, A. (1879). On the vegetable remains at the Egyptian museum at Berlin (edited from the author's manuscript by P. Ascherson and P. Magnus) *J. Bot. N. S.* **8**, 19–23, 48–62, 91–2.

Brown, A. G. (1975). Apples. In: *Advances in fruit breeding* (eds. J. Janick and J. N. Moore), pp. 3–37. Purdue University Press, West Lafayette, Indiana.

Browicz, K. (1972). *Prunus, Amygdalus, Malus, Pyrus.* In: *Flora of Turkey* (ed. P.H. Davis), Vol. 4, pp. 8–12, 21–8, 157–9, 160–8. Edinburgh University Press.

Browicz, K. (1982). *Chorology of trees and shrubs in south-west Asia and adjacent regions.* Vol. 1. 172 pp. Polish Scientific Publishers, Warsaw-Poznań.

Browicz, K. (1986). *Chorology of trees and shrubs in south-west Asia and adjacent regions.* Vol. 5. 88 pp. Polish Scientific publishers, Warsaw-Poznań.

Browicz, K. (1988). *Chorology of trees and shrubs in south-west Asia and adjacent regions.* Vol. 6. 86 pp. Polish Scientific publishers, Warsaw-Poznań.

Browicz, K. (1992). *Chorology of trees and shrubs in south-west Asia and adjacent regions.* Vol. 9. 83pp. Sorus, Poznań.

Browicz, K. and Zieliński, J. (1984). *Chorology of trees and shrubs in south-west Asia and adjacent regions*, Vol. 4. 80 pp. Polish Scientific Publishers, Warsaw-Poznań.

Browicz, K. and Zohary, D. (in press). The genus *Amygdalus* L. (Rosaceae): species, relationships, distribution and evolution under domestications. *Plant Syst. Evol.*

Buschan, G. (1895). *Vorgeschichtliche Botanik der Cultur- und Nutzpflanzen der alten Welt auf Grund prähistorischer Funde.* 268 pp. J. U. Kern Verlag, Breslau.

Buxó i Capdevila, R. (1985). *Dinàmica de l'alimentacio vegetal a partir de l'anàlisi de llavors i fruits.* Tesi de Llicenciatura, Bella-terra, Univ. de Barcelona, Fac. de Lletras. 217 pp.

Cârciumaru, M. (1991). Étude paléobotanique pour les habitats Néolitiques et Énéolithiques de Roumanie. In: Palaeoethnobotany and Archaeology (ed. E. Hajnalová), pp. 61–74. *Acta Interdisciplinaria Archaeologica* (Nitra), Vol. 7.

Castelletti, L. (1972). Contributo alle ricerche paletnobotaniche in Italia. *Ist. Lomb. Accad. Sci. Lett. Rend. Class. Lett.* **106**, 331–74.

Castelletti, (1976). Rapporto preliminare sui resti vegetali macroscopici della serie neoliticobronzo di Pienza (Siena). *Riv. Archeol. dell'antica provincia e diocesi di Como. Fasc.* 156–7, 243–51.

Caton-Thompson, G. and Gardner, E. W. (1934). *The desert Fayum*, Royal Anthropological Institute of Great Britain and Ireland, London.

Chang, T. T. (1989). Domestication and the spread of the cultivated rices. In: *Foraging and farming: the evolution of plant exploitation.* (eds. D. R. Harris and G. C. Hillman), pp. 408–17. Unwin and Hyman, London.

Chowdhury, K. A. and Buth, G. M. (1970). 4500-year-old seed suggest that true cotton is indigenous to Nubia. *Nature, Lond.* **227**, 85–6.

Chowdhury, K. A. and Buth (1971). Cotton seeds from the Neolithic in Egyptian Nubia and the origin of the Old World Cotton. *Biol. J. Linn. Soc.* **3**, 303–13.

Cleuziou, S. and Costantini, L. (1980). Premiers éléments sur l'agriculture protohistorique de l'Arabie Orientale. *Paléorient* **6**, 245–51.

Cleuziou, S. and Costantini, L. (1982). A l'origine des oasis. *Recherche* **13**, 1180–82.

Condit, I. J. (1947). *The fig*. Chronica Botanica, N. S., Plant Science Books 19, XVIII. 222 pp. Waltham, Mass.

Cooper, W. C. and Chapot, H. (1977). Fruit production with special emphasis on fruit for processing. In: *Citrus science and technology* (eds. S. Nagi, P. E. Shur, and M. K. Valdhuis), Vol. 2, pp. 1–127. Avi Publications, Westport, Conn.

Coquillat, M. (1956). Examen des graines du 'Trou Arnaud' près de S. Nazaire-le-Désert (Drôme). *Cahiers Rhodaniens* **3**, 26–32.

Coquillat, M. (1964). Étude paléobotanique et détermination des graines. In: L. Bonnamour: Un habitat protohistorique à Ouroux-sur-Saône (Saône-et-Loire), *Rev. Archéol. Est et du Centre-Est* **15**, 150–3.

Corner, E. H. H. (1966). *The natural history of the palms*, pp. 322–8. Weidenfeld & Nicolson, London.

Costantini, L. (1977). Le piante. In: *La citta bruciata del deserto salato*. The burnt city in the salt desert (ed. G. Tucci), pp. 159–71. Erizzo, Venice-Mestre.

Costantini, L. (1979). Plant remains at Pirak. In: *Fouilles de Pirak* (eds. J.-F. Jarrige and M. Santoni), Vol. 1, pp. 326–33. Boccard, Paris.

Costantini, L. (1984). The beginning of agriculture in the Kachi Plain: The evidence of Mehrgarh. *South Asian archaeology 1981*. Proc. 6th Int. Conf. Assoc. South Asian Archaeologists in western Europe (ed. B. Allchin), pp. 29–33. Cambridge University Press.

Costantini, L. (1989). Plant exploitation at Grotta dell' Uzzo, Sicily: new evidence for the transition from Mesolithic to Neolithic subsistence in southern Europe. In: *Foraging and farming: the evolution of plant exploitation* (eds. D. R. Harris and G. C. Hillman), pp. 197–206. Unwin & Hyman, London.

Costantini, L. and Costantini-Biasini, L. (1985). Agriculture in Baluchistan between the 7th and the 3rd millennium BC. *Newsl. Baluchistan Stud.* (Istituto Universitario Orientale, Naples) **2**, 16–30.

Costantini, L. and Costantini-Biasini, L. (1986). Palaeoethnobotanical investigations in the Middle East and Arabian Peninsula, 1986. *East and West* (Instituto Ital. Med. Estr. Oriente) **36**, 354–64.

Courtin, J. and Erroux, J. (1974). Aperçu sur l'agriculture préhistorique dans le Sud-Est de la France. *Bull. Soc. Préhist. Franç.* **71**, 321–34.

Courtin, J. and Erroux, J. and Thommert, J. H. (1976). Les céréales du Néolithique ancien de Chateauneuf-Les-Martigues (Bouches-du-Rhône). *Bull. Mus. Hist. Nat. Marseille* **36**, 11–15.

Crane, M. B. and Lawrence, W. J. C. (1952). *The genetics of garden plants* (4th edn). Macmillan, London.

Dagan, J. and Zohary, D. (1970). Wild tetraploid wheat from West Iran cytogenetically identical with Israeli *T. dicoccoides*. *Wheat Inform. Serv.* **31**, 15–17.

Darby, W. J., Ghalioungui, P., and Grivetti, L. (1977). *Food: the gift of Osiris*, 2 vols. 876 pp. Academic Press, London.

Davies, D. R., Berry, G. L., Heath, M. C., and Dawkins, T. C. K. (1985). Pea (*Pisum sativum* L.) In: *Grain legume crops* (eds. R. J. Summerfield and E. H. Roberts), pp. 147–98. Collins, London.

Davis, P. H. (1970). *Pisum* L. In: *Flora of Turkey* (ed. P. H. Davis) Vol. 3, pp. 370–3. Edinburgh University Press.

De Candolle, A. (1886). *Origin of cultivated plants* (2nd edition). 468 pp. Reprinted 1964 by Hafner Publ. Comp., New York.

Decker, E., Guillaume, Ch., and Michels, R. (1977). Le gisement rubané linéaire récent du 'Dolem', Kirschnaumen-Evendorff, Moselle. Datation C^{14}. *Bull. Soc. Préhist. Franç.* **74**, 155–60.

De Lattin, B. (1939). Über den Ursprung und die Verbreitung der Reben. *Züchter* **11**, 217–225.

De Wet, J. M. J. (1975). Evolutionary dynamics of cereal domestication. *Bull. Torrey Bot. Club*, **102**, 307–12.

De Wet, J. M. J. (1981). Species concepts and systematics of domesticated cereals. *Kulturpflanze* **29**, 177–98.

De Wet, J. M. J., Harlan, J. R., and Price, E. G. (1976). Variability in *Sorghum bicolor*. In: *The origins of African plant domestication* (eds. J. R. Harlan, J. M. J. De Wet and A. B. L. Stemler), pp. 453–63. Mouton, The Hague.

De Wet, J. M. J., Oestry-Stidd, L. L., and Cubero, J. I. (1979). Origins and evolution of foxtail millets *Setaria italica*. *J. Agric. Trop. Bot. Appl.* **26**, 53–64.

Dörfler, W. (1990). Die Geschichte des Hanfanbaus in Mitteleuropa auf Grund palynologischer Untersuchungen und von Grossrestnachweisen. *Praehistorische Z.* **65**, 218–44.

Dowson, V. H. W. (1982). *Date production and protection*. Tech. Paper No. 35. F.A.O., Rome. 294 pp.

Duke, J. A. (1973). Utilization of *Papaver*. *Econ. Bot.* **27**, 390–400.

Duke, J. A. (1981). *Handbook of legumes of world economic importance*. 345 pp. Plenum Press, New York.

Durrant, A. (1976). Flax and linseed. In: *Evolution of crop plants* (ed. N. W. Simmonds), pp. 190–3. Longman, London.

Eastwood, G. M. (1984). Egyptian dyes and colours. In: *Dyes on historical and archaeological textiles*. 3rd meeting of York Archaeological Trust, September (1984). pp. 9–19.

Einset, J. and Pratt. C. (1975). Grape Vine. In: *Advances in fruit breeding* (eds. J. Janick and J. N. Moore), pp. 130–53. Purdue University Press, West Lafayette, Indiana.

Ellison, R., Renfrew, J., Brothwell, D., and Seeley, N. (1978). Some food offerings from Ur, excavated by Sir Leonard Wooley, and previously unpublished. *J. Archaeol. Sci.* **5**, 167–77.

Epstein, C. (1978). A new aspect of chalcolithic culture. *Bull. Am. Schools Orient. Res.* **229**, 27–45.

Erroux, J. (1976). Les débuts de l'agriculture en France: les céréales. In: *La Préhistoire Française*, Vol. 2, *Les Civilisations Néolithiques et Protohistoriques de la France* (ed. J. Guilaine), pp. 186–91. CNRS, Paris.

Erroux, J. (1979). Détermination des graines carbonisées. In: J. L. Roudil, O. Roudil, and M. Soulier, *La Grotte de l'Aigle à Mejannes-les Clap (Gard)*. **1**, 75.

Evans, A. J . (1928). *The Palace of Minos at Knossos* Vol. 2. London.

Evans, G. M. (1976). Rye. In: *Evolution of crop plants* (ed. N. W. Simmonds), pp. 108–11. Longman, London.

Feinbrunn, N. (1938). New data on some cultivated plants and weeds of early Bronze Age in Palestine. *Palestine J. Botany, Jerusalem Series* **1**, 238–40.

Feldman, M. (1976). Wheats. In: *Evolution of crop plants* (ed. N. W. Simmonds), pp. 120–8. Longman, London.

Feliks, Y. (1983). *Plants and animals of the Mishna*. 343 pp. Institute of Mishna Research, Jerusalem. (In Hebrew.)

Fietz, A. (1936). Prähistorische Pflanzenreste aus der Slowakei. *Verhandl. Naturf. Vereines in Brünn* **67**, 149–51.

Fogle, H. W. (1975). Cherries. In: *Advances in fruit breeding* (eds. J. Janick and J. N. Moore), pp. 348–66. Purdue University Press, West Lafayette, Indiana.

Follieri, M. (1973). Cereali de villaggio neolitico di Passo di Corvo (Foggia). *Annali di Botanica*. **32**, 49–58.

Follieri, M. (1982). Le più antiche testimonianze dell'agricultura neolitica in Italia meridionale. *Rivista 'Origini'. Preistoria e Protostoria delle Civiltà antiche* **11**, 337–44.

Follieri, M. and Castelleti, L. (1987). Palaeobotanical research in Italy. *Quaternario* **1**, 34–9.

Ford-Lloyd, B. V. and Williams, J. J. (1975). A revision of *Beta* section *Vulgares* (Chenopodiaceae), with new light on the origin of cultivated beets. *Bot. J. Linn. Soc.* **71**, 89–102.

Forni, G. (1979). Origini delle strutture agrarie dell'Italia preromana. In: *L'acienda agraria nell'Italia centro-settentrionale dall'antichita ad oggi* (Verona Conference, 1977), pp. 13–66. Giannini, Napoli.

French, D. H., Hillman, G. C., Payne, S., and Payne, R. J. (1972). Excavations at Can Hasan III, 1969–70. In: *Papers in economic prehistory* (ed. E.S. Higgs), pp. 181–90. Cambridge University Press.

Fritsch, R. (1979). Zur Samenmorphologie des Kulturmohns (*Papaver somniferum* L.). *Kulturpflanze* **27**, 217–27.

Galil, J., and Neeman, G. (1977). Pollen transfer and pollination in the common fig (*Ficus carica* L.). *New Phytol.* **79**, 163–71.

Galil, J., Stein, M., and Horovitz, A. (1976). On the origin of the sycamore fig (*Ficus sycomorus* L.) in the Middle East. *Gardens' Bull.* **29**, 191–205.

Galili, E., Weinstein-Evron, M., and Zohary, D. (1989). Appearance of olives in submerged Neolithic sites along the Carmel coast. *Mitekufat Haeven: J. Israel Prehist. Soc.* **22**, 95–7.

Gallant, T. W. (1985). The agronomy, production, and utilization of sesame and linseed in the Graeco-Roman world. *Bulletin on Sumerian agriculture* (eds. J. N. Postgate and M. A. Powell), Vol. 2, pp. 153–8. Cambridge University Press.

Garfinkel, Y., Kislev, M. E., and Zohary, D. (1988). Lentil in the Pre-Pottery Neolithic B Yiftah'el: additional evidence of its early domestication. *Israel J. Bot.* **37**, 49–51.

Germer, R. (1985). *Flora des pharaonischen Ägypten*. 259 pp. Verlag Philipp von Zabern, Mainz.

Germer, R. (1989a). *Die Pflanzenmaterialien aus dem Grab des Tutanchamun*. 94 pp. Gerstenberg, Hildesheim.

Germer, R. (1989b). Ägyptens auswärtige Beziehungen hinsichtlich der Kulturgewächse. *Dissertationes Bot.* **133**, 57–66.

Germer, R. (1992). *Die Textilfärberei und die Verwendung gefärbter Textilien im alten Ägypten*. Ägyptologische Abhandlungen, vol. 53, 150pp. Harrassowitz, Wiesbaden.

Gill, K.S. and Yermanos, D.M. (1967). Cytogenetic studies on the genus *Linum*. I. Hybrids among the taxa with 15 as the haploid chromosome number. *Crop Sci* . **7**, c27–31.

Gillespie, R. (1986). *Radiocarbon user's handbook*. 36 pp. Oxford University Press.

Giżbert, W. (1960). Studium porównawcze nad ziarnami zyta kopalnego (A comparative study on excavated grains of rye). *Archaeol. Polski* **5**, 81–90.

Giżbert, W. (1966). Zbadan nad orkiszem (*Triticum spelta* L.) w Polsce. (Recherches sur l'épautre (Triticum spelta L.) en Pologne). *Materialy Archeologiczne* **7**, 11–18.

Głuża, I. (1983). Neolithic cereals and weeds from the locality of the Lengyel culture at Nowa Huta-Mogiła near Cracow. *Acta Palaeobot. Polonica* **23**, 123–84.

Godwin, H. (1967). The ancient cultivation of hemp. *Antiquity* **41**, 42–8.

Gophna, R. and Kislev, M.E. (1979). Tell Saf (1977–1978). *Revue Biblique* **86**, 112–14.

Grasselly, Ch. and Crossa-Raynaud, P. (1980). *L' Amandier*. 446 pp. G. P. Maisonneuve & Larose, Paris.

Green, P. S. and Wickens, G. E. (1989). The *Olea europaea* complex. In: *The P. H. Davis & I. C. Hedge Festschrift*. (ed. K. Tan). pp. 287–299. Edinburgh University Press.

Greig, J. R. A. (1991). The British Isles. In: *Progress in Old World palaeoethnobotany* (eds. W. van Zeist, K. Wasylikowa and K. -E. Behre), pp. 299–330. Balkema, Rotterdam.

Hajnalová, E. (1978). Funde von *Triticum*-Resten aus einer hallstattzeitlichen Getreide-Speichergrube in Bratislava-Devin/CSSR. *Ber. Dtsch. Bot. Ges.* **91**, 85–96.

Hajnalová, E. (1980). Palaeoethnobotanical findings from the multi-layer Nova Zagora settlement. *Studia Praehistorica* **4**, 91–8. (Russian, with English summary.)

Hajnalová, E. (1989). Katalóg zvyskov semien a plodov v archeologických nálezoch na Slovensku. *Acta Interdisciplinaria Archaeol.* (Nitra, Czechoslovakia) **6**, 3–192.

Halstead, P. and Jones, G. (1980). Early economy in Thessaly. *Anthropologika* **1**, 93–117.

Hammer, K. and Fritsch, R. (1977). Zur Frage nach der Ursprungsart des Kulturmohns (*Papaver somniferum* L.). *Kulturpflanze*, **25**, 113–24.

Hanelt, P. (1963). Monographische Uebersicht der Gattung *Carthamus* (Compositae). *Feddes Repert. Spec. Nov.* **67**, 41–180

Hanelt, P. (1972). Die infraspezifische Variabilität von *Vicia faba* und ihre Gliederung. *Kulturpflanze* **20**, 75–128.

Hanelt, P. (1985). Zur Taxonomie, Chorologie und Ökologie der Wildarten von *Allium* L. sect. *Cepa* (Mill.) Prokh. *Flora* **176**, 99–116.

Hansen, J. M. (1978).The earliest seed remains from Greece: Palaeolithic through Neolithic at Franchthi Cave. *Ber. Dtsch. Bot. Ges.* **91**, 39–46.

Hansen, J. M. (1988). Agriculture in the prehistoric Aegean: data versus speculations. *Am. J. Archaeol.* **92**, 39–52.

Hansen, J. (1991). Palaeobotany in Cyprus: recent research. In: *New light on early farming: recent developments in palaeoethnobotany* (ed. J. M. Renfrew), pp. 225–36. Edinburgh University Press.

Hansen, J. (1992). Franchthi Cave and the beginnings of agriculture in Greece and the Aegean. In: *Préhistoire de l'agriculture: Nouvelles approches expérimentales et ethnographiques* (ed. P. C. Anderson-Gerfaud), pp. 231–247. Monographie du Centre de Recherches Archéologiques No.6. éd. CNRS, Paris.

Harlan, J. R. (1971). Agriculture origins: centers and non-centers. *Science* **174**, 468–74.

Harlan, J. R. (1975). *Crop and man.* 295 pp. American Society for Agronomy, Madison.

Harlan, J. R. (1981). The early history of wheat: earliest traces to the sack of Rome. In: *Wheat science – today and tomorrow* (eds. L. T. Evans and W. J. Peacock), pp. 1–19. Cambridge University Press.

Harlan, J.R. and De Wet, J. M. J. (1971). Towards a rational classification of cultivated plants. *Taxon* **20**, 509–17.

Harlan, J. R. and De Wet, J. M. J. (1973). On the quality of evidence for origin and dispersal of cultivated plants. *Curr. Archaeol.* **14**, 51–62.

Harlan, J. R. and Stemler, A. B. L. (1976). The races of Sorghum in Africa. In: *The origins of African plant domestication* (eds. J. R. Harlan, J. M. J. De Wet, and A. B. L. Stemler), pp. 465–78. Mouton, The Hague.

Harlan, J. R. and Zohary, D. (1966). Distribution of wild wheats and barley. *Science* **153**, 1074–80.

Harlan, J. R. De Wet, J. M. J., and Price, E. G. (1973). Comparative evolution of cereals. *Evolution* **27**, 311–25.

Harris, D. R. (1986). Plant and animal domestication and the origin of agriculture; the contribution of radiocarbon accelerator dating. In: *Archaeological results from accelerator dating* (ed. J. A. J. Gowlett and R. E. M. Hedge). Oxford University Committee for Archaeology Monograph II, pp. 5–21.

Hartyányi, B. P. (1982). A Tiszaalpár-Várdomb bronzkori lakóteleprol származó mag-és termésleletek. (Die von der bronzeseitlichen Wohnsiedlung Tiszaalpár-Várdomb stammenden Korn- und Fruchtfunde). *Cumania*, **7**, *Archaeologia* pp. 133–286.

Hartyányi, B. P. and Máthé, M. Sz. (1979). Pflanzliche Überreste einer Wohnsiedlung aus dem Neolithikum im Karpaten-Becken. *Archaeo-Physika* **8**, 97–114.

Hartyányi, B. P. and Nováki, G. (1975). Samen- und Fruchtfunde in Ungarn von der Neusteinzeit bis zum 18. Jahrhundert. *Agrártörténeti Szemle: Historia Rerum Rusticarum*, **17** (E suppl.), 1–88.

Hedrick, M. P., Howe, G. H., Taylor, O. M., Francis, E. H., and Tukey, H. B. (1921). *The pears of New York.* New York State Dept. of Agriculture 29th annual report, Vol. 2, Part II, 636 pp. Ithaca, NY.

Heer, O. (1865). Die Pflanzen der Pfahlbauten. *Neujahrsblatt der Naturforschenden Gesellschaft Zürich für das Jahr 1866*, **68**, 1–54.

Hegi, G. (1935). *Illustrierte Flora von Mittel-Europa.* Vol. 1 (2nd edn), 528 pp. Hanser Verlag, Munich.

Heim, J. (1979). Recherche paléobotanique au site néolithique (Roessen) de la 'Bosse de la Tombe' à Givry. *Bull. Soc. R. Belge. Anthropol. Préhist.* **90**, 65–78.

Helbaek, H. (1952a). Spelt (*Triticum spelta*) in Bronze Age Denmark. *Acta Archaeol.* **23**, 97–107.

Helbaek, H. (1952b). Preserved apples and Panicum in the prehistoric sites at Nørre Sandegaard in Bornholm. *Acta Archaeol.* **23**, 107–15.

Helbaek, H. (1952c). Early crops in southern England. *Proc. Prehist. Soc.* N.S. **18**, 194–233.

Helbaek, H. (1953). Queen Ichetis' wheat. Kongelige Danske Videnskabernes Selskab. *Biol. Medd.* **21**, (8), 1–17.

Helbaek, H. (1954). Prehistoric food plants and weeds in Denmark. A survey of archaeobotanical research 1923–1954. *Danm. Geol. Unders.* **11**, (fasc. 80), 250–61.

Helbaek, H. (1955). Ancient Egyptian wheats. *Proc. Prehist. Soc.* N.S. **21**, 93–5.

Helbaek, H. (1958). Plant economy in ancient Lachish. In: *Lachish (Tell ed-Duweir) IV. The Bronze Age* (ed. O. Tufnell), Appendix A, pp. 309–17. Oxford University Press.

Helbaek, H. (1959a). Notes on the evolution and history of *Linum*. *Kuml* 103–29.

Helbaek, H. (1959b). Domestication of food plants in the Old World. *Science* **130**, 365–72.

Helbaek, H. (1960a). The paleobotany of the near east and Europe. In: *Prehistoric investigations in Iraqi Kurdistan*. Studies in Ancient Oriental Civilizations No. 31 (eds. R. J. Braidwood and B. Howe), pp. 99–118). University of Chicago Press.

Halbaek, H. (1960b). Ancient crops in the Shahrzoor valley in Iraqi Kurdistan. *Sumer* **16**, 79–81.

Helbaek, H. (1962). Les. grains carbonisés de la 48ème couche des fouilles de Tell Soukas. *Ann. Archéologiques de Syrie* **11–12,** 185–6.

Helbaek, H. (1963). Isin Larsan and Horian food remains of Tell Basmosian in the Dokan Valley. *Sumer* **19**, 27–35.

Halbaek, H. (1964a). First impressions of the Çatal Hüyük plant husbandry. *Anatolian Stud.* **14**, 121–3.

Helbaek, H. (1964b). Early Hassunan vegetable at Es-Sawwan near Samarra. *Sumer* **20**, 45–8.

Helbaek, H. (1966a). Pre-pottery Neolithic farming at Beidha. *Palest. Explor. Quart.* **98**, 61–6.

Helbaek, H. (1966b). The plant remains from Nimrud. In: *Nimrud and its remains* ed. M. E. Mallowan, Appendix I, Vol. 2, pp. 613–20. Collins, London.

Helbaek, H. (1966c). Report on carbonized grain from AF5 (GHD. Phase). Appendix IV. In: D. H. Trump, *Skorba*, Reports of the Research Committee of the Society of Antiquaries of London, Vol. 22, p. 53. Oxford University Press.

Helbaek, H. (1966d). 1966 – Commentary on the phylogenesis of *Triticum* and *Hordeum*. *Econ. Bot.* **20**, 350–60.

Helbaek, H. (1969). Plant collecting, dry-farming and irrigation agriculture in prehistoric Deh Luran. In: *Prehistory and human ecology of the Deh Luran Plain* (eds. F. Hole, K. V. Flannery, and J. A. Neely), pp. 383–426. Memoirs Mus. Anthrop. No. 1, University of Michigan, Ann Arbor.

Helbaek, H. (1970). The plant husbandry of Hacilar. In: *Excavations at Hacilar* (ed. J. Mellaart), Vol. 1 pp. 189–244. Edinburgh University Press.

Helbaek, H. (1972). Samarran irrigation agriculture at Choga Mami in Iraq. *Iraq* **34**, 35–48.

Hepper, F. N. (1990). *Pharaoh's flowers, the botanical treasures of Tutankhamun.* Her Majesty's Stationary Office, London. 80 pp.

Herre, W. and Röhrs, M. (1977). Zoological consideration on the origins of farming and domestication. In: *Origin of agriculture* (ed. C. A. Reed), pp. 245–79). Mouton, The Hague.

Hesse, C. O. (1975). Peaches. In: *Advances in fruit breeding* (eds. J. Janick and J. N. Moore), pp. 285–347. Purdue University Press, West Lafayette, Indiana.

Hillman, G. (1975). The plant remains from Tell Abu Hureyra: A preliminary report. *Proc. Prehist. Soc.* **41**, 70–3.

Hillman, G. (1978). On the origins of domestic rye – *Secale cereale*: the finds from aceramic Can Hasan III in Turkey. *Anatolian Stud.* **28**, 157–74.

Hillman, G. (1984). Interpretation of archaeological plant remains: The application of ethnographic models from Turkey. In: *Plants and ancient man* (eds. W. van Zeist and W. A. Casparie), pp. 1–41. Balkema, Rotterdam.

Hillman, G. C. (1986). Plant foods in ancient diet: The role of palaeofaeces in general and in Lindau man's gut contents in particular. In: *Lindau Man: the body in the bog.* (eds. I. M. Brothwell and J. Bourke), pp. 99–115. British Museum, London.

Hillman, G.C. and Charles, M.P. (in press). Neolithic crop husbandry in a desert environment: preliminary results from charred plant remains from Jeitun. In: Proceedings of the 9th meeting of the international workgroup for palaeoethnobotany, Kiel 1992 (Title of volume pending; ed. H. Kroll).

Hillman, G. C. and Davies, M. S. (1990). Measured domestication rates in wild wheats and barley under primitive cultivation and their archaeological implications. *J. World Prehist.* **4**, 157–222.

Hingst, H. (1973). Eine bronzezeitliche Siedlung bei Schmalstede, Kr. Rendsburg-Eckernförde; mit einem Beitrag von Maria Hopf, Mainz. *Offa* **30**, 194–204.

Hjelmqvist, H. (1955). Die älteste Geschichte der Kulturpflanzen in Schweden. *Opera Bot.* **1**, 1–186.

Hjelmqvist, H. (1963). Zur Geschichte des Einkorns und des Emmers in Schweden. *Bot. Not.* **11**, 487–97.

Hjelmqvist, H. (1969). Getreideabdrücke in den bronzezeitlichen Funden aus Hötofta (Skåne) Schweden. *Acta Archaeol. Lundensia* **8**, 208–16.

Hjelmqvist, H. (1979a). Beiträge zur Kenntnis der prähistorischen Nutzpflanzen in Schweden. *Opera Bot.* **17**, 3–58.

Hjelmqvist, H. (1979b). Some economic plants and weeds from the Bronze Age of Cyprus. *Stud. Mediterranean Archaeol.* **45**(5), 110–13.

Hjelmqvist, H. (1985). Economic plants from two Stone Age settlements in southernmost Scania. *Acta Archaeol.* **54**, 57–63.

Ho, P.-T. (1977). The indigenous origins of Chinese agriculture. In: *Origins of agriculture* (ed. C. A. Reed), pp. 413–83. Mouton, The Hague.

Hofmann, E. (1924). Pflanzenreste der Mondseer Pfahlbauten. *Sitzungsber. Akad. d. Wiss. (Wien), math-nat. Kl.*, Abt. 1, **133**, 379–409.

Holden, J. H. W. (1976). Oats. In: *Evolution of crop plants* (ed. N. W. Simmonds), pp. 86–90. Longman, London.

Hopf, M. (1961a). Untersuchungsbericht über Kornfunde aus Vršnik. *Zbornik na Štipskot Naroden Muzej* **2**, 41–5.

Hopf, M. (1961b). Pflanzenfunde aus Lerna/Argolis. *Züchter* **31**, 239–47.

Hopf, M. (1962). Bericht über die Untersuchungen von Samen und Holzkohlresten von der Argissa-Magula aus den präkeramischen bis mittelbronzezeitlichen Schichten. In: *Die deutschen Ausgrabungen auf der Argissa-Magula in Thessalien* (eds. V. Milojčić, J. Boessneck, and M. Hopf) Vol. 1, pp. 101–10. Bonn.

Hopf, M. (1963). Die Probegrabungen auf der frühbandkeramischen Siedlung bei Eitzun, Kreis Wolfenbüttel. In: *Ausgrabungen und Forschungen in Niedersachsen* (ed. F. Niquet), Vol. 1, pp. 44–74. A. Lax. Hildesheim.

Hopf, M. (1967). Analyse de céréales du niveau III (Chasséen). In: A. Galan, *La station néolithique de la Perte du Cros à Saillac (Lot)*. *Gallia Préhist.* **10**, 70–3.

Hopf, M. (1970). Zur Geschichte der Ackerbohne (*Vicia faba* L.). *Jahrb. Röm.-German. Zentralmus. Mainz.* **17**, 306–22.

Hopf, M. (1971). Vorgeschichtliche Pflanzenreste aus Ostspanien. *Madrider Mitteil.* **12**, 101–14.

Hopf, M. (1973a). Pflanzenfunde aus Nordspanien. Cortes de Navarra – El Soto de Medinilla. *Madrider Mitteil.* **14**, 133–42.

Hopf, M. (1973b). Frühe Kulturpflanzen aus Bulgarien. *Jahrb. Röm.-German. Zentralmus. Mainz* **20**, 1–47.

Hopf, M. (1973c). Apfel (*Malus communis* L.); Aprikose (*Prunus armeniaca* L.). In: *Reallexikon der Germanischen Altertumskunde* (eds. H. Beck, H. Jankuhn, K. Ranke, and R. Wenskus) Vol. 1, pp. 368–72; 375. Walter de Gruyter, Berlin.

Hopf, M. (1974–77). Pflanzenreste aus Siedlungen der Vinča-Kultur in Jugoslawien. *Jahrb. Röm.-German. Zentralmus. Mainz*, **21**, 1–11.

Hopf, M. (1975). Pflanzenfunde von Tell Goljamo Delcevo. *Razkopi i prouĕvanija* **5**, 303–24. (Bulgarian, with Russian and German summary.)

Hopf, M. (1976). Beerenobst; Bier. In: *Reallexikon der Germanischen Altertumskunde* (eds. H. Beck, H. Jankuhn, K. Ranke, and R. Wenskus) Vol. 2, pp. 132–9; 530–3. Walter de Gruyter, Berlin.

Hopf, M. (1978a). Plant remains. In: R. Amiran *et al.*, *Early Arad. I. The Chalcolithic settlement and Early Bronze Age city*, pp. 64–82. Israel Explor. Soc. Jerusalem.

Hopf, M. (1978b). Birne (*Pirus communis* L.); Bohne (*Vicia faba* L.). In: *Reallexikon der Germanischen Alterumskunde* (eds. H. Beck, H. Jankuhn, K. Ranke, and R. Wenskus) Vol. 3, pp. 29–32; 183–7. Walter de Gruyter, Berlin.

Hopf, M. (1981a). Pflanzliche Reste aus Zambujal. In: E. Sangmeister and H. Schubart, *Zambujal, Die Grabungen 1964 bis 1973*, pp. 315–40. Madrider Beiträge Band 5. Deutsches Archäologisches Institut, Madrid. Verlag Philipp von Zabern, Mainz.

Hopf, M. (1981b). Bucheckern. In: *Reallexikon der Germanishcen Altertumskunde* (eds. H. Beck, H. Jankuhn, K. Ranke, and R. Wenskus) Vol. 4, p. 59. Walter de Gruyter, Berlin.

Hopf, M. (1982). *Vor-und frühgeschichtliche Kulturpflanzen aus dem nördlichen*

Deutschland. Kataloge vor-und frühgeschichtlicher Altertümer, 22. 108 pp. *Jahrb. Röm.-German. Zentralmus. Mainz.*

Hopf, M. (1983). Jericho plant remains. In: *Excavations at Jericho*, Vol. 5 (eds. K. M. Kenyon and T. A. Holland), pp. 576–621. British School of Archaeology in Jerusalem, London.

Hopf, M. (1985). Bronzezeitliche Sämereien aus Ouroux-Marnay, Dép. Saône-et-Loire. *Jahrb. Röm.-German. Zentralmus. Mainz* 32, 255–64.

Hopf, M. (1991a). South and southwest Europe. In: *Progress in Old World Palaeoethnobotany* (eds. W. van Zeist, K. Wasylikowa and K.-E. Behre), pp. 241–77. Balkema, Rotterdam.

Hopf, M. (1991b). Kulturpflanzenreste aus der Sammlung Siret in Brüssel. In: H. Schubart and H. Ulreich, *Die Funde der Südostspanischen Bronzezeit aus der Sammlung Siret*, pp. 397–413. Madrider Beiträge Band 17. Philipp von Zabern, Mainz.

Hopf, M. and Bar-Yosef, D. (1987). Plant remains from Hayonim Cave, western Galilee. *Paléorient* 13, 117–20.

Hopf, M. and Blankenhorn, B. (1984). Vor- und frühgeschichtliche Kulturpflanzen in Süddeutschland. *Jahrb. Bayer. Bodendenkmalpflege* 22, (1981).

Hopf, M. and Muñoz, A.M. (1974). Neolithische Pflanzenreste aus der Höhle Los Murciélagos bei Zuheros (Prov. Córdoba). *Madrider Mitteil.* 15, 9–27.

Hopf, M. and Pellicer Catalán, M. (1970). Neolithische Getreidefunde in der Höhle von Nerja (Prov. Málaga). *Madrider Mitteil.* 11, 18–34.

Hopf, M. and Schubart, H. (1965). Getreidefunde aus der Coveta de l'Or (Prov. Alicante). *Madrider Mitteil.* 6, 20–38.

Hopf, M. and Willerding, U. (1989). Pflanzenreste. In: *Bastan II.* (ed. W. Kleiss), pp. 263–318. Mann, Berlin.

Hubbard, R. N. L. (1979). Ancient agriculture and ecology at Servia. Appendix 2 in C. Ridley and K. A. Wardle: Rescue excavations at Servia 1971–1973: a preliminary report. *The Annual of the British School at Athens, London.* 74, 226–228.

Hutchinson, J. B. (1976). India: Local and introduced crops. *Phil. Trans. R. Soc. Lond. B* 275, 129–41.

Jacomet-Engel, S. (1980). Botanische Makroreste aus den neolithischen Seeufersiedlungen des Areals 'Pressehaus Ringier' in Zürich (Schweiz). Stratigraphische und vegetationskundliche Auswertung. *Vierteljahrsschrift der Naturforschenden Gesellschaft in Zürich* 125, 73–163.

Jacomet, S., Brombacher, Ch., and Dick, M. (1989). *Archäobotanik am Zürichsee. Ackerbau, Sammelwirtschaft und Umwelt von neolithischen und bronzezeitlischen Seeufersiedlungen im Raum Zürich.* Züricher Denkmalpflege Monographien. Band 7. Orell Füssli. Zürich. 348 pp.

Jacomet, S., Brombacher, C., and Dick, M. (1991). Palaeobotanical work on Swiss Neolithic and Bronze Age lake dwellings over the past ten years. In: *New light on early farming: recent developments in palaeoethnobotany* (ed. Jane M. Renfrew), pp. 257–76. Edinburgh University Press.

Janushevich, Z. V. (1975). Fossil remains of cultivated plants in the south west Soviet Union. *Folia Quaternaria* 46, 23–30.

Janushevich, Z. V. (1976). Kulturnije rastenija jugu-sapada SSSR po paleobotani-cheskim issledovanijam (Cultivated plants in south western USSR according to paleobotanical investigations). *Akad. Nauk Moldavskoi SSR, Botan. Sad.* pp. 1–213.

Janushevich, Z. V. (1978). Prehistoric food plants in the south-west of the Soviet Union. *Ber. Dtsch. Bot. Ges.* **91**, 59–66.

Janushevich, Z. V. (1984). The specific composition of wheat finds from ancient agricultural centres in the USSR. In: *Plants and ancient man* (eds. W. van Zeist and W. A. Casparie), pp. 267–76. Balkema, Rotterdam.

Jarrige, J.-F. and Meadow, R. H. (1980). The antecedents of civilization in the Indus Valley. *Sci. Am.* **243**(2), 102–10.

Jasny, N. (1944). *The wheats of classical antiquity.* Studies in History and Political Science, Series 62, No.3, 176 pp. Johns Hopkins Press, Baltimore.

Jeffrey, C. (1980). A review of *Cucurbitaceae. Bot. J. Linn. Soc.* **81**, 233–47.

Jennings, D.L. (1976). Raspberries and blackberries. In: *Evolution of crop plants* (ed. N. W. Simmonds). pp. 251–4. Longman, London.

Jensen, H. A. (1991). The Nordic countries. In: *Progress in Old World palaeoethnobotany* (eds. W. van Zeist, K. Wasylikowa, and K.-E. Behre), pp. 335–50. Balkema, Rotterdam.

Jessen, K. (1939). Kornfund. In: *Bundsø, en yngre Stenalders Boplads paa Als* (ed. Th. Mathiassen), pp. 65–84. Aarbøger for Nord. Oldkynd. og Hist.

Jessen, K. and Helbaek, H. (1944). Cereals of Great Britain and Ireland in prehistoric and early historic times. *Kongel. Danske, Vidensk. Selsk. Biol. Skrift* **3**(2), 68 pp.

Jørgensen, G. (1977). Acorns as a food-source in the late stone age. *Acta Archaeol.* **48**, 233–8.

Jørgensen, G. (1979). A new contribution concerning the cultivation of spelt, *Triticum spelta* L., in prehistoric Denmark. *Archaeo-Physica* **8**, 135–45.

Jørgensen, G. (1981). Korn fra Sarup; Med nogle bemaerkinnger om agerbruget i yngre stenalder i Danmark. *Kuml* 221–31.

Kajale, M. D. (1974). Ancient grains from India. *Bull. Deccanian Coll. Res. Inst., Poona* **34**, 55–74.

Kajale, M. D. (1991). Current status of Indian palaeoethnobotany: introduced and indigenous food plants with a discussion of the historical and evolutionary development of Indian agriculture and agricultural systems in general. In: *New light on early farming: recent developments in palaeoethnobotany* (ed. J. M. Renfrew), pp. 155–89. Edinburgh University Press.

Karg, S. and Müller, J. (1990). Neolithische Getreidefunde aus Pokrovnik, Dalmatien. *Archäologisches Korrespondenzblatt* **20**, 373–86.

Keimer, L. (1924). *Die Gartenpflanzen im alten Ägypten.* (1967 reprint) Vol. 1, 117 pp. Hoffman u. Campe, Hamburg/Berlin.

Keimer, L. (1984). *Die Gartenpflanzen im alten Ägypten.* Vol. 2, 86 pp. (Brought to press by R. Germer.) Verlag Philipp von Zabern, Mainz.

Kerber, E. R. and Rowland, G. G. (1974). Origin of the free-threshing character in hexaploid wheat. *Can. J. Genet. Cytol.* **16**, 145–54.

Khush, G. S. (1963a). Cytogenetic and evolutionary studies in *Secale*. III. Cytogenetics of weedy ryes and origin of cultivated rye. *Econ. Bot.* **17**, 60–71.

Khush, G. S. (1963b). Cytogenetic and evolutionary studies in *Secale*. IV. *Secale vavilovii* and its biosystematic status. *Z. Pflanzenzüchtung* **50**, 34–43.

Kihara, H. (1944). Die Entdeckung des DD-analysators beim Weizen. *Agric. Hortic. Japan* **19**, 889–90.

Kislev, M. E. (1980). *Triticum parvicoccum* sp. nov., the oldest naked wheat. *Israel J. Bot.* **28**, 95–107.

Kislev, M. E. (1985). Early Neolithic horsebean from Yiftah'el, Israel. *Science* **228**, 319–20.

Kislev, M. E. (1988). Nahal Hemar cave: desiccated plant remains, an interim report. *'Atiqot* (Dept. Antiquities & Museums, Jerusalem) **38**, 76–81.

Kislev, M. E. (1989). Origins of the cultivation of *Lathyrus sativus* and *L. cicera*. (Fabaceae). *Econ. Bot.* **43**, 262–70.

Kislev, M.E., Bar-Yosef, O., and Gofer, A. (1986). Early Neolithic domesticated and wild barley from Netiv Hagdud region in the Jordan Valley. *Israel J. Bot.* **35**, 197–201.

Kislev, M. E., Nadel, D., and Carmi, I. (1992). Grain and fruit diet 19,000 years old at Ohalo II, Sea of Galilee, Israel. *Rev. Paleobot. Palynol.* **73**, 161–6.

Klichowska, M. (1966). Badania odcisków roślinnych na ceramice z neolitycznego stanowiska w Szlachcinie w pow. średzkim. *Przegląd Archeologiczny* **17**, 84–6.

Klichowska, M. (1976). Aus paläoethnobotanischen Studien über Pflanzenfunde aus dem Neolithikum und der Bronzezeit auf polnischem Boden. *Archaeologia polona* **17**, 27–67.

Knörzer, K.-H. (1971). Prähistorische Mohnsamen im Rheinland. *Bonner Jahrb.* **171**, 34–9.

Knörzer, K.-H. (1973). Der bandkeramische Siedlungsplatz Langweiler 2: Pflanzliche Grossreste. *Rhein. Ausgrab.* **13**, 139–52.

Knörzer, K.-H. (1974). Bandkeramische Pflanzenfunde von Bedburg-Garsdorf, Kreis Bergheim/Erft. *Rhein. Ausgrab.* **15**, 173–92.

Knörzer, K.-H. (1979). Über den Wandel der angebauten Körnerfrüchte und ihrer Unkrautvegetation auf einer niederrheinischen Lössfläche seit dem Frühneolithikum. *Archaeo-Physika* **8**, 147–89.

Knörzer, K.-H. (1991). Deutschland nördlich der Donau. In: *Progress in Old World palaeoethnobotany* (eds. W. van Zeist, K. Wasylikowa, and K.-E. Behre), pp. 189–206. Balekma, Rotterdam.

Knowles, P. F. (1969). Centers of plant diversity and conservation of germ plasm: Safflower. *Econ. Bot.* **23**, 324–9.

König, M. (1989). Ein Fund römerzeitlicher Traubenkerne in Piesport/Mosel. *Dissertationes Botanicae* **133**, 107–116.

Körber-Grohne, U. (1967). *Geobotanische Untersuchungen auf der Feddersen Wierde/Bremerhaven*. 357 pp. Steiner-Verlag, Wiesbaden.

Körber-Grohne, U. (1984). Über die Notwendigkeit einer Registrierung und Dokumentation wilder und primitiver Fruchtbäume, zu deren Erhaltung und zur Gewinnung von Vergleichsmaterial für paläo-ethnobotanische Funde. In: *Plants and ancient man* (eds. W. van Zeist and W. A. Casparie), pp. 237–46. Balkema, Rotterdam.

Körber-Grohne, U. (1987). *Nutzpflanzen in Deutschland*. 490 pp. Konrad Theiss, Stuttgart.

Kranz, A. R. (1961). Cytologische Untersuchungen und genetische Beobachtungen an den Bastarden zwischen *Secale cereale* L. und *Secale vavilovii* Grossh. *Züchter* **31**, 219–25.

Kroll, H. (1975). Ur-und frühgeschichtlicher Ackerbau in Archsum auf Sylt. Eine botanische Grossrestanalyse. Ph.D. Thesis, Christian-Albrecht-Univ., Kiel.

Kroll, H. (1979). Kulturpflanzen aus Dimini. *Archaeo-Physika* **8**, 173–89.

Kroll, H. (1980a). Einige vorgeschichtlichen Vorratsfunde von Kulturpflanzen aus Norddeutschland. *Offa* **37**, 372–83.

Kroll, H. (1980b). Mittelalterlich/frühneuzeitliches Steinobst aus Lübeck. *Lübecker Schriften zur Archäologie und Kulturgeschichte (LSAK)* **3**, 167–73.

Kroll, H. (1981a). Thessalische Kulturpflanzen. *Z. Archäol.* **15**, 97–103.

Kroll, H. (1981b). Mittelneolithisches Getreide aus Dannau. *Offa* **38**, 85–90.

Kroll, H. (1982). Kulturpflanzen von Tiryns. *Archäolog. Anzeiger*, 1982 467–85.

Kroll, H. (1983). Kastanas, Ausgrabungen in einem Siedlungshügel der Bronze- und Eisenzeit Makedoniens 1975–1979. Die Pflanzenfunde. *Prähistorische Archäologie in Südosteuropa*, Vol. 2, pp. 1–176. Verlag Volker Spiess, Berlin.

Kroll, H. (1984). Bronze Age and Iron Age agriculture in Kastanas, Macedonia. In: *Plants and ancient man* (eds. W. van Zeist and W. A. Casparie), pp. 243–6. Balkema, Rotterdam.

Kroll, H. (1990). Saflor von Feudvar, Vojvodina. *Archäologisches Korrespondenzblatt* **20**, 41–6.

Kroll, H. (1991). Südosteuropa. In: *Progress in Old World palaeoethnobotany.* (eds. W. van Zeist, K. Wasylikowa, and K.-E. Behre.), pp. 161–77. Balkema, Rotterdam.

Kühn, F. (1981). Rozbory nálezu polních plodin. ed. *Archeologickyústav ČSAV v Brně Prehled Výzkumu* 1979, 75–9.

Küster, H. (1983). Rekonstruktionsversuche zur neolithischen Landwirtschaft nach botanischen Funden aus Eberdingen-Hochdorf (Kreis Ludwigsburg). *Archäologisches Korrespondenzblatt* **13**, 37–9.

Küster, H. (1991). Mitteleuropa südlich der Donau, einschliesslich Alpenraum. In: *Progress in Old World palaeoethnobotany* (eds. W. van Zeist, K. Wasylikowa, and K.-E. Behre), pp. 179–87. Balkema, Rotterdam.

Ladizinsky, G. (1975). On the origin of the broad bean, *Vicia faba* L. *Israel J. Bot.* **24**, 80–8.

Ladizinsky, G. (1986). A new *Lens* from the Middle-East. *Notes R. Bot. Gard. Edin.* **43**, 489–92.

Ladizinsky, G. and Adler, A. (1976a). The origin of chickpea *Cicer arietinum* L. *Euphytica* **25**, 211–17.

Ladizinsky, G. and Adler, A. (1976b). Genetic relationships among the annual species of *Cicer* L. *Theor. Appl. Genet.* **48**, 196–203.

Ladizinsky, G. and van Oss, H. (1984). Genetic relationship between wild and cultivated *Vicia ervilia*. *Bot. J. Linn. Soc.* **89**, 97–100.

Ladizinsky, G. and Zohary, D. (1971). Notes on species delimitation, species relationships and polyploidy in *Avena* L. *Euphytica* **20**, 380–95.

Ladizinsky, G., Braun, D., Goshen, D., and Muehlbauer, F. J. (1984). The biological species of the genus *Lens* L. *Bot. Gaz.* **145**, 253–61.

Lagerstedt, H. B. (1975). Filberts. In: *Advances in fruit breeding* (eds. J. Janick and J. N. Moore), pp. 456–89. Purdue University Press, West Lafayette, Indiana.

Lauer, J. P., Laurent-Täckholm, V., and Åberg, E. (1950). Les plantes découvertes dans les souterrains de l'enceinte du roi Zoser à Saqqarah (IIIᵉ dyn.). *Bull. Inst. d'Egypte* **32**, 121–47.

Lawes, D. A. (1980). Recent developments in understanding, improvement, and use of *Vicia faba*. In: *Advances in legume science* (eds. R. J. Summerfield and A. H. Bunting), pp. 625–36. Royal Botanic Gardens, Kew, Richmond, UK.

Layne, R. E. C. and Quamme, H. A. (1975). Pears. In: *Advances in fruit breeding* (eds. J. Janick and J.N . Moore), pp. 38–80. Purdue University Press, West Lafayette, Indiana.

Lee, J. (1984). Cotton as a world crop. In: *Cotton* (eds. R. J. Kohel and C. F. Lewis), pp. 1–25. American Society of Agronomy. Madison, Wisconsin.

Lemmens, R.H.M.J. and Wessel-Riemens, P.C. (1991). *Indigofera* L. In: *Plant resources of South-East Asia* No. 3: *dye and tanin-producing plants* (eds. R.H.M.J. Lemmens and N. Wulijarni-Soetjipto), pp. 81–3. Pudoc/Prosea, Wageningen.

Lenz, H. O. (1859). *Botanik der alten Griechen und Römer* (1966 reprint), 776 pp. Sändig oHG., Wiesbaden.

Levadaux, L. (1956). Les populations sauvages et cultivées des *Vitis vinifera* L. *Ann. Amélior. Plantes*. **1**, 59–118.

Li, H. L. (1974). An archaeological and historical account of *Cannabis* in China. *Econ. Bot.* **28**, 437–48.

Lisitsina, G. N. (1978). Main types of ancient farming on the Caucasus, on the basis of a palaeo-ethnobotanical research. *Ber. Dtsch. Bot. Ges.* **91**, 47–57.

Lisitsina, G. N. (1984). The Caucasus – A centre of ancient farming in Eurasia. In: *Plants and ancient man* (eds. W. van Zeist and W. A. Casparie), pp. 285–92. Balkema, Rotterdam.

Lisitsina, G. N. and L. V. Prishchepenko, (1977). *Paleoetnobotanicheskiye nakhodki Kavkaza i Blizhnego Vostoka*. 127pp. 'Nauka', Moskva.

Loeschke, S. (1933). *Denkmäler vom Weinbau aus der Zeit der Römer-Herrschaft an Mosel, Saar, und Ruwer*. 66 pp. Röm. Abt. Dtsch. Weinbaumuseums (Trier).

Lopez, P. (1980). Estudio des semillas prehistoricas algunos yacimientos Españoles. *Trabajos de Prehistoria* **37**, 419–32.

Lundström-Baudais, K. (1984). Palaeoethnobotanical investigation of plant remains from a Neolithic lakeshore site in France: Clairvaux, Station III. In: *Plants and ancient man* (eds. W. van Zeist and W. A. Casparie), pp. 293–305. Balkema, Rotterdam.

Maan, S.S. (1973). Cytoplasmic and cytogenetic relationships among tetraploid *Triticum* species. *Euphytica* **22**, 287–300.

MacKey, J. (1966). Species relationships in *Triticum*. *Proc. 2nd Int. Wheat Genetics Symp.*, Lund 1963. *Hereditas*, Suppl. **2**, 237–76.

Maesen, L. J. G. van der (1972). *Cicer*. A monograph on the genus with special reference to the chickpea (*Cicer arietinum*), its ecology and cultivation. Agricultural University, Wageningen. 342 pp. *Mededelingen Landbouw Hogeschool Wageningen* 72–10.

Malzew, A. I. (1930). Wild and Cultivated Oats, section *Euavena* Griseb. *Bull. appl. Bot. Genet. Plant Breeding (USSR)*, Suppl. 38.

Marinval, P. (1988). *L'alimentation végétale en France du Mésolithigue jusqu'à L'Age du Fer*. 192 pp. Editions du Centre National de la Recherche Scientifique. Paris, Toulouse.

Markgraf, F. (1975). *Camelina sativa* (L.) Crantz. In: *Gustav Hegi's Illustrierte Flora von Mitteleuropa* (2nd edn), Vol. 4(1), pp. 342–5. Verlag Paul Parey, Berlin.

Martin, E. and Murphy, P. (1988). West Row Fen. Suffolk: a Bronze Age fen-edge settlement site. *Antiquity* **62**, 353–8.

Masson, V. and Sarianidi V. I. (1972). Central Asia. Turkmenia before the Achaemenids (transl. and ed. R. Tringham), 219 pp. Thames & Hudson, London.

Mathew, B. (1977). *Crocus sativus* and its allies Iridaceae. *Plant Syst. Evol.* **128**, 89–103.

Mattes, W. (1957). Neue Funde im Heilbronner Raum: Pflanzliche Funde aus vorgeschichtlicher Zeit. *Veröffentlichungen des Historischen Vereins Heilbronn* **22**, 15–39.

McCollum, G. D. (1976). Onion and allies. In: *Evolution of crop plants* (ed. N. W. Simmonds), pp. 186–90. Longman, London.

McCreery, D. W. (1979). Flotation of the Bab edh-Dhra and Numeria plant remains. *Ann. Am. Schools Orient. Res.* **46**, 165–9.

McFadden, E. S. and Sears, E. R. (1946). The origin of *Triticum spelta* and its free-threshing hexaploid relatives. *J. Hered.* **37**, 81–90; 107–16.

Meusel, H., Jäger, E., and Weinert, E. (1965). *Vergleichende Chorologie der zentraleuropäschen Flora*, Vol. 1 (text), 583 pp.; Vol. 2 (maps), 258 pp. Verlag Gustav Fischer, Jena, DDR.

Miller, N.F. (1981). Plant remains from Ville Royale II, Susa. *Cahiers de la Délégation Archéologique Française en Iran* **12**, 137–42.

Miller, N. F. (1984). Some plant remains from Khirokitia, Cyprus: 1977 and 1978 excavations. In: A. Le Brun, *Fouilles récentes a Khirokitia (Chypre) 1977–1981* Vol. 1, pp. 183–89. Recherche sur les Civilisations, mémoire 41. Éditions A.D.P.F., Paris.

Miller, N. F. (1986). Vegetation and landuse; archaeological analysis. In: The Chicago Euphrates archaeological project 1980–1984: an interim report (by G. Algaze *et al.*). *Anatolica* **8**, 84–123.

Miller, N. F. (1991). The Near East. In: *Progress in Old World palaeoethnobotany* (eds. W. van Zeist, K. Wasylikowa, and K.-E. Behre), pp. 133–60. Balkema, Rotterdam.

Miller, T.E. (1987). Systematics and evolution. In: *Wheat breeding* (ed. F.G.H. Lupton), pp. 1–30. Chapman & Hall, London.

Moens, M.-F. and Wetterstrom, W. (1988). The agricultural economy of an Old Kingdom town in Egypt's west delta: insights from the plant remains. *J. Near Eastern Studies* **47**, 159–73.

Moffett, L., Robinson, M. A., and Straker, V. (1989). Cereals, fruits and nuts: charred plant remains from Neolithic sites in England and Wales and the Neolithic economy. In: *The beginning of agriculture* (eds. A. Milles, D. Williams, and N. Gardner), pp. 243–61. BAR International Series #496.

Morris, R. and Sears, E. R. (1967). The cytogenetics of wheat and its relatives. In: *Wheat and wheat improvements* (eds. K. S. Quisenberry and L. P. Reitz), pp. 19–87. American Society for Agronomy, Madison.

Munier, P. (1973). *Le Palmier Dattier*, pp. 243–57. Maisonneuve et Larose, Paris.

Muratova, V. (1931). Common beans, *Vicia faba* L. – A botanical-agronomical monograph. *Bull. Appl. Bot. Genet. Plant Breed. (Leningrad)*, Suppl. **50**, 1–295. (Russian with English summary on pp. 248–95.)

Murray, J. (1970). *The first European agriculture, a study of the osteological and botanical evidence until 2000 BC*, 380 pp. Edinburgh University Press.

Neef, R. (1989). Plants. In: *Picking up the threads: A continuing review of excavations at Deir Alla, Jordan* (eds. G. van der Kooij and M. M. Ibrahim), pp. 30–7. University of Leiden Archaeological Centre.

Neef, R. (1990). Introduction, development and environmental implications of olive cultivation: The evidence from Jordan. In: *Man's role in the shaping of the eastern Mediterranean landscape* (eds. S. Bottema, G. Entjes-Nieborg and W. van Zeist), pp. 295–306. Balkema, Rotterdam.

Nesbitt, M. and Summers. G. D. (1988). Some recent discoveries of millets (*Panicum miliaceum* and *Setaria italica*) at excavations in Turkey and Iran. *Anatolian Stud.* **38**, 85–97.

Netolitzky, F. (1914). Die Hirse aus alten Funden. *Sitzungsber. Kaiserl. Akad.d. Wissensch. (Wien), math.-nat. Kl.* 73, Abt. B, 1–35.

Netolitzky, F. (1935). Kulturpflanzen und Holzreste aus dem prähistorischen Spanien und Portugal. *Buletinul facultăţii de Ştiinţe din Cernăuţi* **9**, 4–8.

Neuweiler, E. (1905). Die prähistorischen Pflanzenreste Mitteleuropas mit besonderer Berücksichtigung der schweizer Funde. *Vierteljahrsschrift der Naturforschenden Gesellschaft in Zürich* **50**, 23–134.

Neuweiler, E. (1935). Nachträge urgeschichtlicher Pflanzen. *Vierteljahrsschrift der Naturforschenden Gesellschaft in Zürich* **80**, 98–122.

Neuweiler, E. (1946). Nachträge II urgeschichtlicher Pflanzen. *Vierteljahrsschrift der Naturforschenden Gesellschaft in Zürich* **91**, 122–36.

Nixon, R. W. (1951). The date palm – 'tree of life' in the subtropical deserts. *Econ. Bot.* **5**, 274–301.

Noy, T., Legge, A. J., and Higgs, E. S. (1973). Recent excavations at Nahal Oren, Israel. *Proc. Prehist. Soc.* **39**, 75–99.

Nürnberg-Krüger, U. (1960a). Cytogenetische Untersuchungen an der Gattung *Secale* L. *Z. Pflanzenzüchtung* **44**, 63–72.

Nürnberg-Krüger, (1960b). Cytogenetische Untersuchungen an *Secale silvestre* Host. 1. Der Bastard mit *Secale cereale* L. *Züchter* **30**, 147–50.

Olmo, H. P. (1976). Grapes. In: *Evolution of crop plants* (ed. N. W. Simmonds), pp. 294–8. Longman, London.

Opravil, E. (1979). Rostlinné zbytky z Mohelnice 1. a 2. *Čas. Slezsk. Muz. (Praha)* A. **28**, 1–13, 97–109.

Opravil, E. (1981). Z histoire lnu b našich zemích a ve střdní Evropě. (Aus der Geschichte des Leines in unseren Ländern und in Mitteleuropa). *Archeologicke rozhledy* **33**, 299–305.

Osborn, D. J. (1968). Notes on medicinal and other uses of plants in Egypt. *Econ. Bot.* **22**, 165–77.

do Paço, A. (1954). Semetes pré-históricas do Castro de Vila Nova de S. Pedro. *Anais Acad. Portug. História* (sér. 2) **5**, 281–359.

Pals, J. -P. (1984). Plant remains from Aartswoud, a Neolithic settlement in the coastal area. In: *Plants and ancient man* (eds. W. van Zeist and W. A. Casparie), pp. 267– 76. Balkema, Rotterdam.

Pals, J. -P. and Voorrips, A. (1979). Seeds, fruits and charcoals from two sites in northern Italy. *Archaeo-Physika* **8**, 217–35.

Pearson, G. W. (1987). How to cope with calibration. *Antiquity* **61**, 98–103.

Percival, J. (1921). *The wheat plant*. 463 pp. Duckworth, London.

Pétrequin, P. (1974). Interprétation d'un habitat néolithique en grotte: le niveau XI de Gonvillars (Haute-Saône). *Bull. Soc. Préhist. Franç.* **71**, études et travaux (2), 489– 534.

Piening, U. (1981). Die verkohlten Kulturpflanzenreste aus den Proben der Cortaillod- und Horgener Kultur. In: *Die neolithischen Ufersiedlungen von Twann. Bd. 14, Botanische Untersuchungen. Ergebnisse der Pollen – und Makrorestanalysen zu Vegetation, Ackerbau und Sammelwirtschaft der Cortaillod und Horgener Siedlungen* (eds. B. Ammann. T. Bollinger, St. Jacomet-Engel. H. Liese-Kleiber, and U. Piening), pp. 69–88. Staatlicher Lehrmittelverlag Bern, Schriftenreihe der Erziehungsdirektion – des Kantons Bern, herausgegeben vom Archaeologischen Dienst des Kantons Bern.

Pinto da Silva, A. R. (1976). Carbonized grains and plant imprints in ceramics from the Castrum of Baiões (Beira Alta, Portugal). *Folia Quaternaria* **47**, 3–18.

Pinto da Silva, A. R. (1988). A paleoetnobotanica na arqueologica portuguesa. Resultados desde 1931 a 1987. In: *Paleoecologia e Arqueologia* (eds. F.M.V.R. Queiroga, I.M.A.R. Sousa, and C.M. Oliveira), pp. 5–36. Camara Municipal de Vila Nova de Famalicao.

Postgate, J. N. (1980). Palm-trees, reeds and rushes in Iraq ancient and modern. In: *L'archéologie de L'Iraq: Perspectives et limites de l'interpretation anthropologique des documents*, (ed. M.-Th. Barrelet), pp. 99–110. Colloques internationaux du C.N.R.S. No. 580. France.

Postgate, J. N. (1985). The 'oil plant' in Assyria. *Bulletin on Sumerian agriculture* (eds. J. N. Postgate and M. A. Powell), Vol. 2, pp. 145–52. Cambridge University Press.

Postgate, J. N. (1987). Notes on fruits in the cuneiform sources. *Bulletin on Sumerian agriculture* (eds. J. N. Postgate and M. A. Powell), Vol. 3, pp. 115–44. Cambridge Univ. Press.

Pruessner, A. H. (1920). Date culture in ancient Babylonia. *Am. J. Semitic Languages and Literature* **36**, 213–32.

Punzi, Q. (1968). Le stazioni preistoriche costiere del Brindisino. *Rivista di scienze preistoriche* **23**, 206–21.

Rao, K. E. P., de Wet J. M. J., Brink, D. E., and Mengesha, M. H. (1987). Infraspecific variation and systematics of cultivated *Setaria italica* foxtail millet (*Poaceae*). *Econ. Bot.* **41,** 108– 16.

Rao, P. S. and Smith, E. L. (1968). Studies with Israeli and Turkish accessions of *Triticum turgidum* L. emend. var. *dicoccoides* Körn. Bowden. *Wheat Inform. Serv.* **26**, 6–7.

Rehder, A. (1967). *Manual of cultivated trees and shrubs* (2nd edn), 996 pp. Macmillan, New York.

Renfrew, J. M. (1966). A report on recent finds of carbonized cereal grains and seeds from prehistoric Thessaly. *Thessalika* **5**, 21–36.

Renfrew, J. M. (1968). A note on the Neolithic grain from Can Hasan. *Anatolian Stud.* **18**, 55–6.

Renfrew, J. M. (1972). The plant remains. In: P. Warren: *Myrtos: an early Bronze Age settlement in Crete*. Suppl. 7, pp. 315–17. British School at Athens, London.

Renfrew, J. M. (1973). *Palaeoethnobotany. The prehistoric food plants of the Near East and Europe*. 248 pp. Methuen, London.

Renfrew J. M. (1974). Report on carbonized grains and seeds from Obre I, Kakanj and Obre II. *Wiss. Mitt. Bosnisch-Herzegow. Landesmus*. **4A**, 47–53.

Renfrew, J. M. (1976). Carbonized seed from Anza. In: *Neolithic Macedonia as reflected in the excavation of Anza, south-east Yugoslavia* (ed. M. Gimbutas). Monumenta Archaeologica 1, pp. 300–12. University of California, Los Angeles.

Renfrew, J. M. (1979). The first farmers in South East Europe. *Archaeo-Physika* **8**, 243–65.

Renfrew, J. M. (1984). Cereals cultivated in ancient Iraq. *Bulletin on Sumerian agriculture* (eds. J. N. Postgate and M. A. Powell) Vol. 1, pp. 32–44. Cambridge University Press.

Renfrew, J. M. (1985a). Finds of sesame and linseed in ancient Iraq. *Bulletin on Sumerian agriculture* (eds. J. N. Postgate and M. A. Powell) Vol. 2, pp. 63–6. Cambridge University Press.

Renfrew, J. M. (1985b). Pulses recorded from ancient Iraq. *Bulletin on Sumerian Agriculture* (eds. J. N. Postgate and M. A. Powell), Vol. 2, pp. 67–71. Cambridge University Press.

Reuveni, O. (1986). Fruit set and fruit development of the date palm (*Phoenix dactylifera* L.). In: *Handbook of flowering* (ed. A. Halevi), pp. 119–44. CRC Press, Boca Raton.

Richardson, I. B. K. (1981). Chestnuts (genus *Castanea*). In: *The Oxford encyclopaedia of trees of the world* (ed. B. Hora), pp. 133–5. Oxford University Press.

Riley, R. (1965). Cytogenetics and evolution of wheat. In: *Crop plant evolution* (ed. J. B. Hutchinson), pp. 103–22. Cambridge University Press.

Riviera Núñez, D. and Walker, J. (1989). A review of palaeobotanical findings of early *Vitis* in the Mediterranean and of the origin of cultivated grape-vines, with special reference to new pointers to prehistoric exploitation in the Western Mediterranean. *Rev. Palaeobot. Palynol*. **61**, 205–17.

Rollefson, G. O. and Simmons, A. H. (1985). Excavations at 'Ain Ghazal 1984: preliminary report. *Annual of the Department of Antiquities of Jordan* **29**, 11–30.

Roshevitz, R. I. (1947). Monograph of the genus *Secale* L. *Acta instituti botanici academiae scientiarum USSR* **1** (6), 105–63. (Russian, with English summary.)

Rothmaler, W. and Natho, I. (1957). Bandkeramische Kulturflanzenreste aus Thüringen und Sachsen. *Beitr. Frühgesch. d. Landwirtsch*. **3**, 73–98.

Rubzov, G. A. (1944). Geographic distribution of the genus *Pyrus* and trends and factors in its evolution. *Am. Naturalist* **78**, 358–66.

Schäfer, H. I. (1973). Zur Taxonomie der *Vicia narbonensis*-Gruppe. *Kulturpflanze* **21**, 211–73.

Schick, T. (1988). Nahal Hemar cave: cordage, basketry and fabrics. *'Atiqot* (Dept. Antiquities & Museums, Jerusalem) **38**, 31–43.

Schiel, V. (1913). De l'exploitation des dattiers dans l'ancienne Babylonie. *Revue d'Assyriologie et d'Archéologie Orientale* **10**, 1–9.

Schiemann, E. (1933). Auf den Spuren der ältesten Kulturpflanzen. *Forschungen und Fortschritte Nachrichtenblatt d. Deutschen Wissenschaft und Technik* No. 28, pp. 412–14.

Schiemann, E. (1948). *Weizen, Roggen, Gerste. Systematik, Geschichte, und Verwendung.* 102 pp. Fischer, Jena.

Schiemann, E. (1956). Fünfzig Jahre *Triticum dicoccoides. Ber. Dtsch. Bot. Ges.* **69**, 309–22.

Schlichtherle, H. (1977/1978). Vorläufiger Bericht über die archäobotanischen Untersuchungen an Demircihüyük (Nordwestanatolien). In: M. Korfmann, *Demircihüyük,* pp. 45–53. Istambuler Mitteilungen. Deutsches Archäologisches Institut, Istambul.

Schlichtherle, H. (1981). Cruciferen als Nutzpflanzen in neolithischen Ufersiedlungen Südwestdeutschlands und der Schweiz. *Z. Archäol.* **15**, 113–24.

Schoch, W. H., Pawlik, H., and Schweingruber, F. H. (1988). *Botanische Makroreste.* Paul Haupt, Bern, 227 pp.

Scholz, H. (1983). Die Unkraut-Hirse (*Panicum miliaceum* subsp. *ruderale*) – neue Tatsachen und Befunde. *Plant Syst. Evol.* **143**, 233–44.

Schultze-Motel, J. (1968–84). Literatur über archäologische Kulturpflanzenreste. For 1965–67: *Kulturpflanze* **16**, 215–30 (1968). For 1968: *Jahresschrift für mitteldeutsche Vorgeschichte* **55**, 55–63 (1971). For 1969: *Kulturpflanze* **19**, 265–82 (1972). For 1970/1: *Kulturpflanze* **20**, 191–207 (1972). For 1971/2: *Kulturpflanze* **21**, 61–76 (1973). For 1972/3: *Kulturpflanze* **22**, 61–76 (1974). For 1973/4: *Kulturpflanze* **23**, 189–205 (1975). For 1974/5: *Kulturpflanze* **24**, 159–78 (1976). For 1975/6: *Kulturpflanze* **25**, 71–88 (1977). For 1976/7: *Kulturpflanze* **26**, 349–62 (1978). For 1977/8: *Kulturpflanze,* **27**, 229–45 (1979). For 1978/9: *Kulturpflanze* **28**, 361–78 (1980). For 1979/80: *Kulturpflanze* **29**, 447–63 (1981). For 1980/1: *Kulturpflanze* **30**, 255–72 (1982). For 1981/2: *Kulturpflanze* **31**, 281–97 (1983). For 1982/83: *Kulturpflanze* **32**, 229–43 (1984). For 1983/84: *Kulturpflanze* **33**, 287–305 (1985). For 1984/85: *Kulturpflanze* **34**, 317–33 (1986). For 1985/86: *Kulturpflanze* **35**, 401–20 (1987). For 1986/87: *Kulturpflanze* **36**, 549–69 (1988). For 1987/88: *Kulturpflanze* **37**, 427–51 (1989). For 1988/89: *Kulturpflanze* **38**, 387–416 (1990). For 1989/90: *Veget. Hist. Archaebot.* **1**, 53–62 (1992).

Schultze-Motel, J. (1979a). Die urgeschichtlichen Reste des Schlafmohns (*Papaver somniferum* L.) und die Enstehung der Art. *Kulturpflanze* **27**, 207–15.

Schultze-Motel, J. (1979b). Die Anbaugeschichte des Leindotters: *Camelina sativa* (L.) Crantz. *Archaeo-Physika* **8**, 267–81.

Schultze-Motel, J. (1988). Archäologische Kulturpflanzenreste aus der Georgischen SSR (Teil 1). *Kulturpflanze* **36**, 421–35.

Schultze-Motel, J. (1989). Archäologische Kulturpflanzenreste aus der Georgischen SSR (Teil 2). *Kulturpflanze* **37**, 415–26.

Schultze-Motel, J. and Kruse, J. (1965). Spelz (*Triticum spelta* L.), andere Kulturpflanzen und Unkräuter in der frühen Eisenzeit Mitteldeutschlands. *Kulturpflanze* **13**, 586–619.

Schweinfurth, G. (1891). Aegyptens auswärtige Beziehungen hinsichtlich der Kulturgewächse. *Z. Ethnologie (Organ der Berliner Gesellschaft für Anthropologie, Ethnologie, und Urgeschichte)* **23**, 649–69.

Schweingruber, F. H. (1979). Wildäpfel und prähistorische Äpfel. *Archaeo-Physika* **8**, 283–94.

Sears, E. R. (1969). Wheat cytogenetics. *Ann. Rev. Genet.* **3**, 451–68.

Sencer, H. A. and Hawkes, J. G. (1980). On the origin of cultivated rye. *Biol. J. Linn. Soc. Lond.* **13**, 299–313.

Simmonds, N. W. (1976). Hemp. In: *Evolution of crop plants* (ed. N. W. Simmonds), pp. 203–4. Longman, London.

Širjaev, G. (1932). *Generis Trigonella L. Revisio critica V. Sectio Foenum graecum.* 42 pp. Publ. Fac. Sci. Univ. Masaryk, Brno.

Small, E. and Cronquist, A. (1976). A practical and natural taxonomy for *Cannabis. Taxon.* **25**, 405–35.

Smartt, J. (1990). *Grain lequmes: evolution and genetic resources.* Cambridge University Press, Cambridge. 379 pp.

Smartt, J. and Hymowitz, T. (1985). Domestication and evolution of grain legumes. In: *Grain legume crops* (eds. R. J. Summerfield and E. H. Roberts), pp. 37–72. Collins, London.

Smithson, J. B., Thompson, J. A., and Summerfield, R. G. (1985). Chickpea (*Cicer arietinum* L.). In: *Grain legume crops* (eds. R. J. Summerfield and E. H. Roberts), pp. 312–90. Collins, London.

Snogerup, S. (1980). The wild forms of the *Brassica oleracea* group (2n = 18) and their possible relations to cultivated ones. In: *Brassica crops and wild allies* (eds. S. Tsunada, K. Hinata, and C. Gomez-Campo), pp. 121–32. Japan Scientific Societies Press, Tokyo.

Solheim, W. G. (1972). An earlier agricultural evolution. *Sci. Am.* **226**(6), 34–41.

Spiegel-Roy, P. and Kochba, J. (1981). Inheritance of nut and kernel traits in almond (*Prunus amygdalus* Batsch). *Euphytica* **30**, 167–74.

Stager, L. E. (1985). First fruits of civilization. In: *Palestine in the Bronze and Iron Age: papers in honour of Olga Tufnell* (ed. J. N. Tubb), pp. 172–87. Institute of Archaeology, London.

Stemler, A. B. L. (1980). Origins of plant domestication in the Sahara and the Nile Valley. In: *The Sahara and the Nile* (eds. M. A. J. Williams and H. Faure), pp. 503–26. Balkema, Rotterdam.

Stearn, W. T. (1978). European species of *Allium* and allied genera of *Alliaceae*: a synonymic enumeration. *Ann. Musei Goulandris* **4**, 83–198.

Stewart, R. T. (1974). Paleobotanical investigation: 1972 season. In: *American expedition to Idalion, Cyprus* (eds. L. Stager, A. Walker, and G. E. Wright). *Bull. Am. Schools Orient. Res.* Suppl. 18, pp. 123–9.

Stika, H.-P. (1988). Botanische Untersuchungen in der bronzezeitlichen Höhensiedlung Fuente Alamo. *Madrider Mitteil.* **29**, 21–76.

Stika, H.-P. & Frank, K.-S. (1988). Die Kirschpflaume: Systematik, Morphologie, Verwendung, Genetik und Archäologische Funde. In: *Der prähistorische Mensch und seine Umwelt*: Festschrift U. Körber-Grohne (ed. H. Küster), pp. 65–71. Konard Theiss, Stuttgart.

Stol, M. (1987a). Garlic, onion, leek. *Bulletin on Sumerian agriculture.* (eds. J. N. Postgate and M. A. Powell), Vol. 3, pp. 57–80. Cambridge University Press.

Stol, M. (1987b). The Cucurbitaceae in the cuneiform texts. *Bulletin on Sumerian agriculture* (eds. J.N. Postgate and M.A. Powell), Vol. 3, pp. 81–92.

Storey, W. B. (1976). Fig. In: *Evolution of crop plants* (ed. N. W. Simmonds), pp. 205–8. Longman, London.

Storey, W. B. and Condit, I. J. (1969). Fig (*Ficus carica*). In: *Outlines of perennial crop breeding in the tropics* (eds. F. P. Werda and F. Wit), pp. 259–67. Wageningen, The Netherlands.

Stutz, H. C. (1972). On the origin of cultivated rye. *Am. J. Bot.* **59**, 59–70.

Täckholm, V. (1961). Botanical identification of the plants found in the Monastry of Phoebammon. In: *Le Monastère de Phoebammon dans le Thébaid Tombe III*, pp. 1–38. Publications de la Société d'Archéologie Copte. Rapport des Fouilles, Cairo.

Täckholm, V. (1976). *Faraos Blomster*. 378 pp. Generalstabens Litografiska, Trelleborg, Sweden.

Täckholm, V. and Drar, M. (1950). *Flora of Egypt* Vol. II, *Cyperus esculentus*, pp. 60–9; *Phoenix dactylifera*, pp. 165–273. Bull. Fac. Science No. 28, Fouad University Press, Cairo.

Täckholm, V. and Drar, M. (1954). *Flora of Egypt*, Vol. III, *Allium in ancient Egypt*, pp. 93–106. Cairo University Press, Cairo.

Täckholm, V., Täckholm, G., and Drar, M. (1941). *Flora of Egypt*, Vol. I, *Wheats*, pp. 225–68; Barley, pp. 278–301. Bull. Fac. Science No. 17, Fouad University Press, Cairo.

Takahashi, R. (1955). The origin and evolution of cultivated barley. *Adv. Genet.* **7**, 227–66.

Tanaka, M. and Ishii, H. (1973). Cytogenetic evidence on the speciation of wild tetraploid wheats collected in Iraq, Turkey and Iran. *Proc. 4th Int. Wheat Genet. Symp.* pp. 115–21. University of Missouri.

Tanaka, M., Kawahara, T., and Sano, J. (1979). The origin and the differentiation of the B and G genomes of the tetraploid wheats. *Report Plant Germ-plasm Inst., Kyoto Univ.* **4**, 1–11.

Tempír, Z. (1964). Beiträge zur ältesten Geschichte des Pflanzenbaus in Ungarn. *Acta Archaeologica Academiae Scientiarum Hungaricae* **16**, 65–98.

Tempir, Z. (1966). Výsledky paleoetnobotanického studia pěstování zemědělských rostlin na území ČSSR (Results of paleoethnobotanic studies on the cultivation of agricultural plants in the CSSR). *Vědecké práce Čsl. Zemělského muzea (Praha)* 1966, 27–144.

Tempír, Z. (1968). Archeologické nálezy zemědělských rostlin a prevelů v Čechách a na Moravě (Archaeological finds of agricultural plants and their weeds in Bohemia and Moravia). *Vědecké práce Čsl. Zemělského muzea. (Praha)* 1968, 15–88.

Tempír, Z. (1969). Archeologické nálezy zemědělských rostlin a plevelů na Slovensku (Archaeological finds of food plants and weeds in Slovakia). *Agrikultúra* **8**, 7–66.

Tempír, Z. (1973). Nálezy pravěkých a středověkých zbytků pěstovaných a užitkových rostlin a plevelů na některých lokalitách v Čechách a na Moravě (Finds of prehistoric and early historic remains of food plants and weeds in some sites in Bohemia and Moravia). *Vědecké práce Čsl. Zemělského muzea. (Praha)* 1973, 19–47.

Tempír, Z. (1979). Kulturpflanzen im Neolithikum und Äneolithikum auf dem Gebiet von Böhmen und Mähren. *Archaeo-Physika* **8**, 302–8.

Thiebault, S. (1989). A note on the ancient vegetation of Baluchistan based on charcoal analysis of the latest periods from Mehrgarh, Pakistan. *South*

Asian Archaeology 1985. Papers from the 8th Int. Conf. Assoc. South Asian Archaeologists in Western Europe (eds. K. Frifelt and P. Sorensen), pp. 186–8. Curzon Press, London.

Thompson, R. C. (1949). *A dictionary of Assyrian botany.* 405 pp. The British Academy, London.

Tosi, M. (1975). Hasanlu project 1974: Paleobotanical survey. *Iran* (British Inst. Persian Studies) **8**, 185–6.

Uerpmann, H.-P. (1989). The ancient distribution of ungulate mammals in the Middle East. *Beihefte zum Tübinger Atlas des Vorden Orients*, Reihe A, No. 27. Ludwig Reichert, Wiesbaden.

Vavilov, N. I. (1949–50). Phytogeographic basis of plant breeding. In: Selected writings of N. I. Vavilov: The origin, variation, immunity, and breeding of cultivated plants. *Chronica Bot.* **16** (1/6), 46–54.

Vences, F.J., Vaquero, F., Gracia, P., and Perez de la Vega, M. (1987). Further studies on phylogenetic relationships in *Secale*: on the origin of its species. *Plant Breeding* **98**, 281–91.

Villaret-von Rochow, M. (1958). Die Pflanzenreste der bronzezeitlichen Pfahlbauten von Valeggio am Mincio. *Bericht über das Geobot. Forschungsinst. Rübel in Zürich für das Jahr 1957*, 96–144.

Villaret-von Rochow, M. (1967). Frucht- und Samenreste aus der neolithischen Station Seeberg, Burgäschisee-Süd. *Acta Bernensia* **2**, 21–64.

Villaret-von Rochow, M. (1971). *Avena ludoviciana* Dur. im Schweizer Spätneolithikum, ein Beitrag zur Abstammung des Saathafers. *Ber. Dtsch. Bot. Ges.* **84**, 243–8.

Vishnu-Mittre (1977). Changing economy in ancient India. In: *Origins of agriculture* (ed. C. A. Reed), pp. 569–88. Mouton, The Hague.

Vishnu-Mittre and Savithri, R. (1982). Food economy of the Harappans. In: *Harappan civilization: a contemporary prospective.* (ed. G. L. Possehl). pp. 205–21. Oxford and IBH Publishing Co., New Delhi.

Vuorela, I. and Lempiäinen, T. (1988). Archaeobotany of the site of the oldest cereal grain find in Finland. *Ann. Bot. Fennici* **25**, 33–45.

Waines, J.G. and Barnhart, D. (1992). Biosystematic research in *Aegilops* and *Triticum*. *Hereditas* **116**, 207–12.

Waines, J. G. and Stanley Price, N. P. (1975–77). Plant remains from Khirokitia in Cyprus. *Paléorient* **3**, 281–4.

Warburg, O. (1905). Die Gattung *Ficus* im nichttropischen Vorderasien. In: *Festschrift Paul Ascherson* (eds. U. Ign and P. Graebner), pp. 364–70. Gebr. Borntraeger, Leipzig.

Wasylikowa, K. (1984). Fossil evidence for ancient food plants in Poland. In: *Plants and ancient man* (eds. W. van Zeist and W. A. Casparie), pp. 257–66. Balkema, Rotterdam.

Wasylikowa, K., Cărciumaru, M., Hajnalová, E., Hartyáni B. P., Pashkevich, G. A., and Yanushevich, Z. V. (1991). East-central Europe. In: *Progress in Old World palaeoethnobotany* (eds. W. van Zeist, K. Wasylikowa and K.-E. Behre), pp. 207–39. Balkema, Rotterdam.

Watkins, R. (1981). Apples (genus *Malus*), pears (genus *Pyrus*), and plums, apricots, almonds, peaches, cherries, (genus *Prunus*). In: *The Oxford encyclopaedia of trees of the world* (ed. B. Hora), pp. 187–201. Oxford University Press.

Watson, W. (1969). Early cereal cultivation in China. In: *The domestication and exploitation of plants and animals* (eds. P. J. Ucko and G. W. Dimbleby), pp. 397–402. Duckworth, London.

Webb, D. A. (1968). *Prunus*, pp. 77–80; *Vitis*, p. 246. In: *Flora Europaea*, Vol. 2 (eds. T. G. Tutin, V. H. Heywood, N. A. Burges, D. M. Moore. D. H. Valentine, S. M. Walters, and D. A. Webb). Cambridge University Press.

Weber, S. A. (1991). *Plants and Harappan subsistance*. Oxford & IBH Publishing Co., New Delhi. 200 pp.

Weinberger, J. H. (1975). Plums. In: *Advances in fruit breeding* (eds. J. Janick and J. N. Moore), pp. 336–47. Purdue University Press, West Lafayette, Indiana.

Werneck, H. L. (1949). Ur-und frühgeschichtliche Kultur-und Nutzpflanzen in den Ostalpen und am Rande des Böhmerwaldes. *Schriftenreihe O.-Ö. Landesbaudirektion*, Nr. 6, 288 pp.

Werneck, H. L. (1961). Ur- und frühgeschichtliche sowie mittelalterliche Kulturpflanzen und Hölzer aus den Ostalpen und dem südlichen Böhmerwald (Nachtrag 1949–1960). *Archaeologia Austriaca* **30**, 68–117.

Werneck, H. L. and Bertsch, K. (1959). Zur Ur- und Frühgeschichte der Pflaumen im oberen Rhein- und Donauraume. *Angew. Botanik* **33**, 19–33.

Wetterstrom, W. (1984). The plant remains. In: *Archaeological investigations at El Hibeh 1980; preliminary report* (ed. R.J. Wenke), pp. 50–79. Undena Publications, Malibu, Calif.

White, K. D. (1970). *Roman farming*. 536 pp. Cornell University Press, Ithaca.

Willerding, U. (1970). Vor- und frühgeschichtliche Kulturpflanzen in Mitteleuropa. *Neue Ausgrabungen und Forschungen in Niedersachsen*, Vol. 5, pp. 287–375. A. Lax, Hildesheim.

Willerding, U. (1980). Zum Ackerbau der Bandkeramiker. In: *Beiträge zur Archäologie Nordwestdeutschlands und Mitteleuropas*. Materialhefte zur Ur- und Frühgeschichte Niedersachsens, Vol. 16, pp. 421–56. A. Lax, Hildesheim.

Woenig, F. (1886). *Die Pflanzen im alten Aegypten*. 425 pp. Wilhelm Friedrich, Leipzig.

Wright, T. W. (1981). Hazels (genus *Corylus*). In: *The Oxford encyclopaedia of the trees of the world* (ed. B. Hora), pp. 143–4. Oxford University Press.

Zaprjagaeva, V. I. (1964). *Dikorastuščie plodovye Tadžikistana (Wild growing fruits of Tadzhikistan)*. 'Nauka', Moskva – Leningrad. 695 pp.

van Zeist, W. (1968–70). Prehistoric and early historic food plants in the Netherlands. *Palaeohistoria* **14**, 41–173.

van Zeist, W. (1970). The Oriental Institute excavations at Mureybit, Syria: Preliminary report on the 1965 campaign. Part III. Palaeobotany. *J. Near East. Stud.* **29**, 167–76.

van Zeist, W. (1972). Palaeobotanical results in the 1970 season at Çayönü, Turkey. *Helinium* **12**, 3–19.

van Zeist, W. (1975). Preliminary report on the botany of Gomolava. *J. Archaeol. Sci.* **2**, 315–25.

van Zeist, W. (1976). On macroscopic traces of food plants in southwestern Asia (with some references to pollen data). *Phil. Trans. R. Soc. Lond., Biol. Sci.* **275**, 27–41.

van Zeist, W. (1979–80). Plant remains from Girikihaciyan, Turkey. *Anatolica*, **7**, 75–89.

van Zeist, W. (1980). Aperçu sur la diffusion des végétaux cultivés dans la région Méditerranéene. In: *Colloque sur la mise en place, l'évolution et la caractérisation de la flore et la végétation circumméditerranéenne* (Fundation L. Emberger). Naturalia Monspeliensia, Special volume 1980, pp. 129–45. Montpellier.

van Zeist, W. (1981). Plant remains from Cape Andreas-Kastros (Cyprus). In: *Un Site Néolithique Précéramique en Chypre: Cap Andreas-Kastros* (ed. A. Le Brun), Append. VI, pp. 95–100. Mém. no. 5, Recherche sur les Grandes Civilisations. Editions ADPF, Paris.

van Zeist, W. (1983). Fruit in foundation deposits of two temples. *J. Archaeol. Sci.* **10**, 351–4.

van Zeist, W. (1988). Some aspects of early Neolithic plant husbandry in the Near East. *Anatolica* **15**, 49–67.

van Zeist, W. (1991). Economic aspects. In: *Progress in Old World palaeoethnobotany* (eds. W. van Zeist, K. Wasylikowa, and K.-E. Behre), pp. 109–30. Balkema, Rotterdam.

van Zeist, W. and Bakker-Heeres (1971). Plant husbandry in early Neolithic Nea Nikomedeia, Greece. *Acta Bot. Neerl.* **20**, 524–38.

van Zeist, W. and Bakker-Heeres, J. A. H. (1975a). Evidence for linseed cultivation before 6000 BC. *J. Archaeol. Sci.* **2**, 215–19.

van Zeist, W. and Bakker-Heeres, J. A. H. (1975b). Prehistoric and early historic plant husbandry in the Altinova plain, south-eastern Turkey. In: *Korucutepe. Final report on the excavations of the universities of Chicago, California (Los Angeles) and Amsterdam in the Keban Reservoir, Eastern Anatolia, 1968–70*, Vol. 1 (ed. M. N. van Loon), pp. 223–57. North Holland, Amsterdam.

van Zeist, W. and Bakker-Heeres, J. A. H. (1985). Archaeological studies in the Levant 1. Neolithic sites in the Damascus Basin: Aswad, Ghoraifé, Ramad. *Palaeohistoria* **24**, 165–256.

van Zeist, W. and Bakker-Heeres, J. A. H. (1986a). Archaeological studies in the Levant 2. Neolithic and Halaf levels at Ras Shamra. *Palaeohistoria* **26**, 151–70.

van Zeist, W. and Bakker-Heeres, J. A. H. (1986b). Archaeobotanical studies in the Levant 3. Late-Palaeolithic Mureybit. *Palaeohistoria* **26**, 171–99.

van Zeist, W. and Buitenhuis, H. (1983). A palaeobotanical study of Neolithic Erbaba, Turkey, *Anatolica* **10**, 47–89.

van Zeist, W. and Casparie, W. A. (1968). Wild einkorn wheat and barley from Tell Mureybit in northern Syria. *Acta Bot. Neerl.* **17**, 44–53.

van Zeist, W. and Casparie, W. A. (1974). Niederwil, a palaeobotanical study of a Swiss Neolithic lake shore settlement. *Geologie en Mijnbouw* **53**, 415–28.

van Zeist, W. and Vynckler, J. (1984). Palaeobotanical investigations of Tell ed-Der. In: *Tell ed-Der IV* (ed. L. de Meyer), pp. 119–43. Uitgeverij Peeters, Leuven.

van Zeist, W. and Waterbolk-van Rooijen, W. (1985). The palaeobotany of Tell Bouqras, eastern Syria. *Paléorient* **11**, 131–47.

Yanushevich, Z. V. (1989). Agriculture evolution north of the Black Sea from the Neolithic to the Iron Age. In: *Foraging and farming: the evolution of plant exploitation*. (eds. D. R. Harris and G. C. Hillman) pp. 607–19. Unwin & Hyman, London.

Zhukovsky, P. M. (1964). Kulturnie rastenia i ich sorodiczy (2nd edn). 791 pp. 'Kolos', Leningrad.

Zohary, D. (1965). Colonizer species in the wheat group. In: *The genetics of colonizer species* (eds. H. G. Baker and G. L. Stebbins), pp. 403–23. Academic Press, New York.

Zohary, D. (1969). The progenitors of wheat and barley in relation to domestication and agriculture dispersal in the Old World. In: *The domestication and exploitation of plants and animals* (eds. P.J. Ucko and G. W. Dimbleby), pp. 47–66. Duckworth, London.

Zohary, D. (1970). Centres of diversity and centres of origin: In: *Genetic resources in plants– their exploration and conservation* (eds. D. H. Frankel and E. Bennett), pp. 33–42. Int. Biol. Programmes, London.

Zohary, D. (1971). Origin of southwest Asiatic cereals: wheats, barley, oats, and rye. In: *Plant life of south-west Asia* (eds. P. H. Davis, P.T. Harper, and I. Hedge), pp. 235–60. Botanical Society of Edinburgh.

Zohary, D. (1972). The wild progenitor and the place of origin of the cultivated lentil: *Lens culinaris*. Medik. *Econ. Bot.* **26**, 326–32.

Zohary, D. (1973). The origin of cultivated cereals and pulses in the Near East. In: *Chromosomes today*, Vol. 4 (eds. J. Wahrman and K. R. Lewis), pp. 307–20. John Wiley, New York.

Zohary, D. (1977). Comments on the origin of cultivated broad bean, *Vicia faba* L. *Israel J. Bot.* **26**, 39–40.

Zohary, D. (1983). Wild genetic resources of crops in Israel. *Israel J. Bot.* **32**, 97–127.

Zohary, D. (1984). Modes of evolution in plants under domestication. In: *Plant biosystematics* (ed. W.F. Grant), pp. 579–86. Academic Press Canada, Montreal.

Zohary, D. (1988). Pulse domestication and cereal domestication: how different are they? *Econ. Bot.* **43**, 31–4.

Zohary, D. (1989). Domestication of southwest Asian Neolithic crop assemblage of cereals, pulses, and flax: the evidence from the living plants. In: *Foraging and farming: the evolution of plant exploitation* (eds. D. R. Harris and G. C. Hillman), pp. 358–73. Unwin & Hyman, London.

Zohary, D. (1991). The wild genetic resources of lettuce (*Lactuca sativa* L.) *Euphytica* **53**, 31–5.

Zohary, D. (1992). Is the European plum, *Prunus domestica* L., a *P. cerasifera* Ehrh. × *P. spinosa* L. allopolyploid? *Euphytica* **60**, 75–77.

Zohary, D. and Hopf, M. (1973). Domestication of pulses in the Old World. *Science* **182**, 887–94.

Zohary, D. and Plitman, U. (1979). Chromosome polymorphism, hybridization and colonization in the *Vicia sativa* group (*Fabaceae*). *Plant Syst. Evol.* **131**, 143–56.

Zohary, D. and Spiegel-Roy, P. (1975). Beginnings of fruit growing in the Old World. *Science* **187**, 319–27.

Zohary, D., Harlan, J. R., and Vardi, A. (1969). The wild diploid progenitors of wheat and their breeding value. *Euphytica* **18**, 58–65.

Zohary M. (1972). *Flora Palaestina*, Vol. 2, plates. Israel Academy of Sciences and Humanities, Jerusalem.

Zohary M. (1973). *Geobotanical foundations of the Middle East*, 2 vols. 739 pp. Fischer, Stuttgart.

Supplementary References

This supplement adds several relevant books and papers which were not listed in the 1993 hardcover version of the second edition. Most of these works appeared very recently; and were not available to the authors when they ended the second edition. Wherever space allowed, references to these additional publications were also inserted in the text of the paperback version.

Anderson. P.C. (ed.). (1992). *Préhistoire de l'agriculture: nouvelles approches expérimental et ethnographiques*. Monographie du Centre de Recherces Archeologiques No. 6, C.N.R.S., Paris.

Browicz, K. (1993). Conspect and chorology of the genus *Pyrus* L. *Arboretum Kornickie* (Institute of Dendrology, Kornik, Poland) **38**, 17–33.

Gept, P. (1993). The use of molecular and biochemical markers in crop evolution studies. In: *Evolutionary biology* vol. 27 (ed. M.K. Hecht *et al.*), pp. 51–94. Plenum Press, New York.

Körber-Grohne, U. (1991). The determination of fibre plants in textiles, cordage and wickerwork. In: *New light on early farming* (ed. J. Renfrew), pp. 93–104. Edinburgh University Press.

Lagudah, E.S., Apples, R., McNeil, D., and Schachtman, D.P. (1993). Exploiting the diploid D genome chromatin for wheat improvement. In: *Gene conservation and exploitation. 20th Stadler genetics symposium* (ed. J.P. Gustafson, R. Apples, and P. Raven), pp. 87–107. Plenum Press, New York.

Moore, J.N. and Ballington, J.R. (eds.). (1991). *Genetic resources of temperate fruit and nut crops* (2 volumes). International Society for Horticultural Science, Wageningen, The Netherlands. [Particularly the treatment of: pears (by R.L. Bell); almonds (by D.E. Kester, T.M. Gradziel, and C. Grasselly); walnuts (by G. McGranahan and C. Leslie); hazelnuts (by S.A. Mehlenbacher); and chestnuts (by P. A. Rutter, G. Miller, and J.A. Payne)].

Mullins, M.G., Bouqet, A., and Williams, L.E. (1992). *Biology of the grapevine*. Cambridge University Press.

Index